OMPHALOS:

AN ATTEMPT

TO UNTIE THE GEOLOGICAL KNOT.

BY

PHILIP HENRY GOSSE, F.R.S.

WITH FIFTY-SIX ILLUSTRATIONS ON WOOD.

Αὐξάνεται δὲ τὰ ζῶα πάντα, ὅσα ἔχει ὀμφαλὸν, διὰ τοῦ ὀμφαλοῦ.
ARIST.; *Hist. Anim.* vii. 8.

OX BOW PRESS
Woodbridge, Connecticut

1998 reprint by
Ox Bow Press
P.O. Box 4045
Woodbridge, CT 06525
203 387-5900
fax 203 387-0035

Originally published by John Van Voorst
Paternoster Row, London, 1857

Originally printed by R. Clay
Bread Street Hill, London

Library of Congress Cataloging in Publication Data
Gosse, Philip Henry, 1810–1888.
Omphalos : an attempt to untie the geological knot /
by Philip Henry Gosse.
p.   cm. — (Originally published: London : J. Van Voorst,
1857.)
"With fifty-six illustrations on wood."
ISBN 1-881987-10-8 (bpk. : alk. paper)
1. Evolution (Biology)   2. Creation.   3. Bible and geology.
I. Title.   II. Series.
QH363.G66   1998
576.8′2—dc21   97-45526
                                                          CIP

Printed in the United States of America

PREFACE.

"You have not allowed for the wind, Hubert,"
said Locksley, in "Ivanhoe;" "or that had been
a better shot."

I remember, when I was in Newfoundland, some
five-and-twenty years ago, the disastrous wreck
of the brig *Elizabeth*, which belonged to the
firm in which I was a clerk. The master had
made a good observation the day before, which
had determined his latitude some miles north of
Cape St. Francis. A thick fog coming on, he
sailed boldly by compass, knowing that, according
to his latitude, he could well weather that promontory. But lo! about midnight the ship plunged
right against the cliffs of Ferryland, thirty miles

to the south, crushing in her bows to the wind-
lass; and presently went down, the crew barely
saving their lives. The captain *had not allowed
for the polar current*, which was setting, like a
sluice, to the southward, between the Grand Bank
and the land.

When it was satisfactorily ascertained that the
heavenly body, now known as Uranus, was a
planet, its normal path was soon laid down according to the recognised law of gravitation. But
it would not take this path. There were devia-
tions and anomalies in its observed course, which
could in nowise be referred to the operation of
any known principle. Astronomers were sorely
puzzled to explain the irregularities, and to reconcile facts with laws. Various hypotheses were
proposed: some denied the facts; that is, the
observed places of the planet, boldly assuming
that the observers had been in error: others suggested that perhaps the physical laws, which had
been supposed to govern the whole celestial
machinery, did not reach so far as Uranus's orbit.

The secret is now known : *they had not allowed for the disturbances produced by Neptune.*

In each of these cases the conclusions were legitimately deduced from the recognised premises. Hubert's skilled eye had calculated the distance ; his experience had taught him the requisite angle at which to shoot, the exact amount of force necessary, and every other element proper to insure the desired result, *except one.* There was an element which he had overlooked; and it spoiled his calculations. *He had forgotten the wind.*

The master of the ill-fated brig had calculated his latitude correctly ; he knew the rate of his vessel's speed; the compass had showed him the parallel on which to steer. These premises ought to have secured a safe conclusion ; and so they would, but for an unrecognised power that vitiated all ; he was not aware of the silent and secret current, that was every hour setting him to the south of his supposed latitude.

The path of Uranus had been calculated by the astronomers with scrupulous care, and every known element of disturbance had been considered; not by one, but by many. But for the fact that the planet had been previously seen in positions quite inconsistent with such a path, it would have been set down as beyond controversy correct. Stubborn fact, however, would not give way; and hence the dilemma, till Le Verrier suggested the unseen antagonist.

I venture to suggest in the following pages an element, hitherto overlooked, which disturbs the conclusions of geologists respecting the antiquity of the earth. Their calculations are sound on the recognised premises; *but they have not allowed for the Law of Prochronism in Creation.*

The enunciation of this principle will lie in a nut-shell; the reader will find it at p. 124; or p. 347. All the rest of the book is illustration.

I do not claim originality for the thought which I have here endeavoured to work out. It was suggested to me by a Tract, which I met with some dozen years ago, or more; the title of which I have forgotten: I am pretty sure it was anonymous, but it was published by Campbell, of 1, Warwick Square. Whether it is still in print

I do not know; I never saw another copy. If the author is alive, and if he should happen to cast his eye on this volume, he will doubtless recognise his own bantling, and accept this my acknowledgment.

The germ of the argument, however, I have found, since these pages were written, in "The Mineral and Mosaical Geologies," of Granville Penn (1822). The state of physical science when he wrote did not enable him to press the argument to a demonstration, as I have endeavoured to do; for he could not refer to structural peculiarities as sensible records of past processes, *inseparable from newly created organisms.*

I would not be considered as an opponent of geologists; but rather as a co-searcher with them after that which they value as highly as I do, TRUTH. The path which I have pursued has led me to a conclusion at variance with theirs. I have a right to expect that it be weighed; let it not be imputed to vanity if I hope that it may be accepted.

But what I much more ardently desire is, that

the thousands of thinking persons, who are scarcely satisfied with the extant reconciliations of Scriptural statements and Geological deductions,—who are silenced but not convinced,—may find, in the principle set forth in this volume, a stable resting-place. I have written it in the constant prayer that the God of Truth will deign so to use it; and if He do, to Him be all the glory!

P. H. G.

MARYCHURCH, TORQUAY,
October, 1857.

CONTENTS.

I.

THE CAUSE.

Evidence of the Senses often delusive—Deductions of Reason fallible—Essentials sometimes overlooked—Discrepancy between Scripture and Geological Conclusions—Painful Dilemma —Efforts to escape from it—Supremacy of Truth—Various Attempts at Reconciliation—Denouncers—Opinions of Brown — Blackwood—Macbrair—Ure—Penn— Young—Cockburn— Miller — Sedgwick — Turner— Sumner—Chalmers—Harris— Gray—Conybeare—Hitchcock—Pye Smith—" Protoplast "— Babbage — Powell — " Vestiges " — Amplitude of Choice

II.

THE WITNESS FOR THE MACRO-CHRONOLOGY.

A Court of Inquiry—The Witnesses—Testimony of One—Strata of Thames Tunnel—of Hertfordshire—of Yorkshire—of the Globe—Granite—Granitic Strata—Organic Remains—Silurian System—Corals—Trilobites—Mollusks—Devonian System — Old Red Sandstone—Its Formation—Fishes—Carboniferous System—Coral Limestone—Millstone Grit—Coal—Predominance of Carbonic Acid—Extent and Thickness of Coal-Fields —Formation of Coal—Conjecture as to its Age—Antediluvian Theory untenable—Sauroid Fishes—Earliest Reptiles—Footprints of Frogs . . . .

III.

THE SAME—(*continued.*)

IV.

THE CROSS-EXAMINATION.

V.

POSTULATES.

VI.

LAWS.

VII.

PARALLELS AND PRECEDENTS.

(*Plants.*)

VIII.

PARALLELS AND PRECEDENTS.

(*Invertebrate Animals.*)

IX.

PARALLELS AND PRECEDENTS.

(*Vertebrate Animals.*)

X.

PARALLELS AND PRECEDENTS.

(*Man.*)

XI.

PARALLELS AND PRECEDENTS.

(*Germs.*)

XII.

THE CONCLUSION.

b

LIST OF ILLUSTRATIONS.

Ὁ ʾΟΜΦΑΛΟΣ.

I.

THE CAUSE.

"Is there not a cause?"—1 Sam. xvii. 29.

An eminent philosopher has observed that
" nothing can be more common or frequent than
to appeal to the evidence of the senses as the most
unerring test of physical effects. It is by the
organs of sense, and by these alone, that we can
acquire any knowledge of the qualities of external
objects, and of their mutual effects when brought
to act one upon another, whether mechanically,
physically, or chemically ; and it might, therefore,
not unreasonably be supposed, that what is called
the evidence of the senses must be admitted to be
conclusive, as to all the phenomena developed by
such reciprocal action. .

B

" Nevertheless, the fallacies are numberless into which those are led who take what they consider the immediate results of sensible impressions, without submitting them to the severe control and disciplined analysis of the understanding." *

If this verdict is confessedly true with regard to many observations which we make on things immediately present to our senses, much more likely is it to be true with respect to conclusions which are not " the immediate results of sensible impressions," but are merely deduced by a process of reasoning from such impressions. And if the direct evidence of our senses is to be received with a prudent reserve, because of this possibility of error, even when we have no evidence of an opposing character, still more necessary is the exercise of caution in judging of facts assumed to have occurred at a period far removed from our own experience, and which stand in contradiction (at least apparent, *primâ facie*, contradiction) to credible historic testimony. Nay, the caveat acquires a greatly intensified force, when the testimony with which the assumed facts are, or seem to be, at variance, is no less a testimony than His who ordained the " facts," who made the objects of

* Dr. Lardner; Museum of Science and Art, vol. i. p. 81.

investigation; the testimony of the Creator of all things; the testimony of Him who is, from eternity to eternity, "῾Ο ἈΨΕΥΔΗΣ ΘΕΟΣ"!

I hope I shall not be deemed censorious in stating my fear that those who cultivate the physical sciences are not always sufficiently mindful of the "*Humanum est errare.*" What we have investigated with no little labour and patience, what we have seen with our eyes many many times, in many aspects, and under many circumstances, we naturally believe firmly; and we are very prone to attach the same assurance of certainty to the inferences we have, *bonâ fide*, and with scrupulous care to eliminate error, deduced from our observations, as to the observations themselves; and we are apt to forget that some element of error may have crept into our actual investigations, and still more probably into our deductions. Even if our observations be so simple, so patent, so numerous, as *almost* to preclude the possibility of mistake in them, and our process of reasoning from them be without a flaw, still we may have overlooked a principle, which, though perhaps not very obvious, ought to enter into the investigation, and which, if recognised, would greatly modify our conclusions.

In this volume I venture to suggest such a principle to the consideration of geologists. It will not be denied that Geology is a science that stands peculiarly in need of being cultivated with that salutary self-distrust that I have above alluded to. Though a strong and healthy child, it is as yet but an infant. The objects on which its senses have been exercised, its τὰ βλεπόμενα, are indeed plain enough and numerous enough, when once discovered; but the inferences drawn from them, its βεβαία, find their sphere in the most venerably remote antiquity,—an antiquity mensurable not by years or centuries, but by *secula seculorum.* And the dicta, which its votaries rest on as certitudes, are at variance with the simple literal sense of the words of God.

I am not assuming here that the Inspired Word has been rightly read; I merely say that the plain straightforward meaning, the meaning that lies manifestly on the face of the passages in question, is in opposition with the conclusions which geologists have formed, as to the antiquity and the genesis of the globe on which we live.

Perhaps the simple, superficial sense of the Word is not the correct one; but it is at least that which its readers, learned and unlearned, had been generally content with before ; and which would, I suppose, scarcely have been questioned, but for what appeared the exigencies of geological facts.

Now while there are, unhappily, not a few infidels, professed or concealed, who eagerly seize on any apparent discrepancy between the works and the Word of God, in order that they may invalidate the truth of the latter, there are, especially in this country, many names of the highest rank in physical (and, among other branches, in geological) science, to whom the veracity of God is as dear as life. They cannot bear to see it impugned ; they know that it cannot be overthrown ; they are assured that He who gave the Word, and He who made the worlds, is One Jehovah, who cannot be inconsistent with Himself. But they cannot shut their eyes to the startling fact, that the records which *seem* legibly written on His created works do flatly contradict the statements which *seem* to be plainly expressed in His word.

Here is a dilemma. A most painful one to the reverent mind! And many reverent minds have laboured hard and long to escape from it. It is unfair and dishonest to class our men of science with the infidel and atheist. They did not rejoice in the dilemma ; they saw it at first dimly, and

hoped to avoid it.* At first they believed that
the mighty processes which are recorded on
the " everlasting mountains " might not only be
harmonized with, but might afford beautiful and
convincing demonstrations of Holy Scripture. They
thought that the deluge of Noah would explain
the stratification, and the antediluvian era account
for the organic fossils.

As the " stone book " was further read, this
mode of explanation appeared to many untenable;
and they retracted their adherence to it. To a
mind rightly constituted, Truth is above every
thing : there is no such thing as a pious fraud ;
the very idea is an impious lie : God is light, and
in Him is no darkness at all ; and that religion
which can be maintained only by dissembling or
denying truth, cannot proceed from " Him that is

---

* As Cuvier, Buckland, and many others. On the question
whether the phenomena of Geology can be comprised within the
short period formerly assigned to them, the Rev. Samuel Charles
Wilks long ago observed : "Buckland, Sedgwick, Faber, Chalmers,
Conybeare, and many other Christian geologists, strove long
with themselves to believe that they could: and they did not
give up the hope, or seek for a new interpretation of the sacred
text, till they considered themselves driven from their position
by such facts as we have stated. If, *even now, a reasonable, or
we might say* POSSIBLE *solution were offered, they would,* we feel
persuaded, *gladly revert* to their original opinion."—*Christian
Observer*, August, 1834.

Holy, Him that is True," but from him who " is a liar, and the father of it."

Many upright and ardent cultivators of the young science felt that truth would be compromised by a persistence in those explanations which had hitherto passed current. The discrepancy between the readings in Science and the hitherto unchallenged readings in Scripture, became manifest. Partisans began to array themselves on either side; some, jealous for the honour of God, knew little of science, and rushed into the field ill-prepared for the conflict; some, jealous for science, but little conversant with Scripture, and caring less for it, were willing to throw overboard its authority altogether: others, who knew that the writings were from the same Hand, knew therefore that there must be some way of reconciling them, and set themselves to find it out.

Have they succeeded? If I thought so, I would not publish this book. Many, I doubt not, have been convinced by each of the schemes by which the discrepant statements have been sought to be harmonized. Each of them has had sufficient plausibility to convince its propounder; and, probably, others too. And some of them have attained a large measure of public confidence. Yet if any one of them is true, it certainly has not commanded uni-

versal assent. Let us examine how far they agree
among themselves, who propose to reconcile Scripture and Science, "the Mosaic and the Mineral
Geologies."

And first, it is, perhaps, right to represent the
opinions of those who stand by the literal accepta-
tion of the Divine Word. There have been some,
indeed, who refuse to entertain the question of reconciliation, taking the high ground that, as the
Word of God is and must be true, it is impious to
set any evidence in competition with it. I cannot
but say, my sympathies are far more with these
than with those who, at the opposite pole of the
argument, would make scientific deduction para-
mount, and make the Word go to the wall. But,
then, we ought to be quite sure that we have got
the very Word of God; and, so far from being impious, it seems highly proper and right, when conflicting evidence appears to flow out of what is
indubitably God's *work*, to examine afresh the
witnesses on both sides, that we may not make
either testify what it does not.

Those good men who merely *denounce* Geology
and geologists, I do not quote. There are the facts,
"written and engraven in stones," and that by the
finger of God. How can they be accounted for?
Some have recourse to the assumption that the

natural processes by which changes in the earth's surface are now going on, may have operated in antediluvian times with a rapidity and power of which we can form little conception from what we are cognisant of. The Rev. J. Mellor Brown takes this ground, adducing the analogies of steam-power and electricity, as effecting in a few moments or hours, what formerly would have required several days or weeks to accomplish.

" God's most tremendous agencies may have been employed in the beginning of his works. If, for instance, it should be conceded that the granitic or basaltic strata were once in a state of fusion, there is no reason why we should not call in the aid of supposition to produce a *rapid* refrigeration. We may surround the globe with an atmosphere (not as yet warmed by the rays of the newly kindled sun) more intensely cold than that of Saturn. The degree of cold may have been such as to cool down the liquid granite and basalt in a few hours, and render it congenial to animal and vegetable life; while the gelid air around the globe may have been mollified by the abstracted caloric."*

A writer in Blackwood (xli. 181; xlii. 690), in like manner, adheres to the literal sense of Genesis

* Reflections on Geology.

and the Decalogue, and alludes to "the great agencies—the magnetic, electrical, and ethereal influences—probably instrumental in all the phenomena of nature," as being far more powerful than is generally suspected.

Mr. Macbrair—who does not, however, appear, from the amount of his acquaintance with science, competent to judge of the physical evidence—supposes stratification to have proceeded with immense rapidity, because limestone is now deposited in some waters at the rate of six inches per annum. Because a mass of timber, ten miles in length, was collected in the Mississippi, in thirty-eight years, he considers that a "capital coal field" might be formed in a single century. Alluvial strata are mud lavas ejected from volcanoes. The whole difficulty of fossil remains is got rid of by ignoring the distinctions of species, and assuming that the ancient animals and the recent ones are identical. The Pterodactyle and the Plesiosaurus he does not allude to.*

According to Dr. Ure,—"The demiurgic week ... is manifestly composed of six working days like our own, and a day of rest, each of equal length, and, therefore, containing an evening and a morn-

* Geology and Geologists.

ing, measured by the rotation of the earth round its axis ... Neither reason nor revelation will justify us in extending the origin of the material system beyond six thousand years from our own days. The world then received its substance, form, and motions from the volition of the Omnipotent.''

His theory of the stratification extends over the whole antediluvian era. He supposes that successive irruptions of the central heat broke up the primitive strata and deposited the secondary and tertiary. "The basaltic or trap phenomena lead to the ·conclusion that such upheavings and subversions were not confined to one epoch of the antediluvian world, but that, coeval with its birth, they pervaded the whole period of its duration ... The Deluge—that universal transflux of the ocean—was the last and greatest of these terraqueous convulsions." *

Another class of this school of interpreters refers the stratification of the earth, either to the deluge alone, or to that convulsion conjoined with the one which is considered to have taken place on the third day of the Mosaic narrative. Perhaps the most eminent writer of this class is Mr. Granville Penn, whose opinions may be thus condensed.

* New System of Geology.

He supposes that this globe has undergone only two revolutions. The first was the violent rupture and depression of the surface to become the bed of the sea, and the simultaneous elevation of the other portion to become dry land,—the theatre of terrestrial existence. This first revolution took place before the creation of any organized beings. The second revolution was at the Noachic Flood, when the former bed of the sea was elevated to become the dry land, with all its organic accumulations of sixteen centuries, while the former land was correspondingly depressed and overflowed. " The earth must, therefore, necessarily exhibit manifest and universal evidences of the vast apparent ruin occasioned by its first violent disruption and depression; of the presence and operation of the marine fluid, during the long interval which succeeded; and of the action and effects of that fluid in its ultimate retreat."*

Mr. Fairholme † so nearly agrees with the above, that I need not quote his opinions in detail.

Another class, represented by Dr. Young and the Rev. Sir W. Cockburn, Dean of York, have maintained with considerable power, backed by no mean geological knowledge, that the deluge is a sufficient

* Mineral and Mosaic Geologies, p. 430. † Geology of Scripture.

vera causa for the stratification of the globe, and for the fossilization of the organic remains.

Dr. Young supposes that an equable climate prevailed all over the globe in the antediluvian period. " Were the highest mountains transferred to the equatorial regions, the most extensive oceans removed towards the poles, and fringed with a border of archipelago,—while lands of moderate height occupied most of the intermediate spaces, between these archipelagos and the equatorial mountains ; then a temperature, almost uniform, would prevail throughout the world." This "perpetual summer" would account for the prodigious quantities of animal and vegetable remains :—every region teemed with life.

At the Flood, " the bed of the ocean must have been elevated, and the dry land at the same time depressed," an expansive force acting from below to heave up the ocean's bed. To this agency are attributed the vast masses of granite, gneiss, basalt, and other rocks of igneous origin, which seem to have been forced upwards in a state of fusion, into their present lofty stations. The ancient bed of the ocean may have consisted of numerous layers of sand, clay, lime, and other substances, including corals and marine shells,—to a certain degree consolidated into rocks. By the progressive rising of the waters and the currents so made, fresh materials would be conveyed to the depths of the ocean, so that the magnesian limestone, the saliferous beds, the lias, &c., would be deposited.*

The Dean of York, in like manner, considers that the convulsions produced by the Deluge, are sufficient to account for all the stratification and fossil remains. That the gradual rise of the waters, and their penetration into the recesses of the rocks, would cause successive volcanic eruptions; the earlier of which would inclose marine fishes and reptiles; then others in turn, the pachyderms and great reptiles of the plains; and, finally, the creatures more exclusively terrestrial. That these repeated heavings of mighty volcanoes raised great part of what had been the bottom of the sea, above its level, and that hence the present land had been for sixteen centuries under water. That the animals which entered the ark, were not selected till after many species had already perished in the earlier convulsions, and hence the number of extinct species now exhumed.†

My reader will kindly bear in mind, that I am not examining these opinions; I adduce them as

* Scriptural Geology, *passim.*   † Letter to Buckland, 15, *et seq.*

examples of the diversity of judgment that still prevails on a question which some affect to consider as settled beyond the approach of doubt.

A totally different solution of the difficulty has been sought in the hypothesis, that the six "days" of the Inspired Record signify six successive periods of immense though of undefined duration. This opinion is as old as the Fathers at least,* and not a few able maintainers of it belong to our own times. It has been put forth, however, with most power, by a late lamented geologist, whose wonderful vigour of description and felicity of illustration, have done, perhaps, more than the efforts of any other living man, to render his favourite science popular.

Perhaps I can scarcely set his views in a more striking light than he himself has done in his own peculiarly graphic report of a conversation, which he sustained with some humble inquirers in the Paleontological Gallery of the British Museum.

" I last passed," says Mr. Hugh Miller, " through this wonderful gallery at the time when the attraction of the Great Exhibition had filled London with curious visitors from all parts of the empire; and a group of intelligent mechanics, fresh from

* Origen, Augustine, &c.

some manufacturing town in the midland counties,
were sauntering on through its chambers immediately before me. They stood amazed beneath
the dragons of the Oolite and Lias; and, with more
than the admiration and wonder of the disciples of
old, when contemplating the huge stones of the
Temple, they turned to say, in almost the old words,
' Lo! master, what manner of great beasts are
these?' 'These are,' I replied, ' the sea-monsters
and creeping things of the second great period of
organic existence.' The reply seemed satisfactory,
and we passed on together to the terminal apart-
ments of the range appropriated to the tertiary
organisms. And there, before the enormous mammals, the mechanics again stood in wonder, and
turned to inquire. Anticipating the query, I said,
' And these are the huge beasts of the earth, and
the cattle of the third great period of organic
existence ; and yonder in the same apartment, you
see, but at its farther end, is the famous fossil Man
of Guadaloupe, locked up by the petrifactive
agencies in a slab of limestone.' The mechanics
again seemed satisfied; and, of course, had I en-
countered them in the first chamber of the suite,
and had they questioned me respecting the organisms with which *it* is occupied, I would have told

them that they were the remains of the herbs and trees of the *first* great period of organic existence. But in the chamber of the mammals we parted, and I saw them no more." *

A large and influential section of the students of Geology regard this hypothesis as untenable. Generally they may be described as holding that the history which is recorded in the igneous and fossiliferous strata does not come into the sacred narrative in any shape. As, however, that narrative commences with " the beginning," and comes down to historic times, the facts so recorded must find their chronology within its bounds. Their place is accordingly fixed by this school of· interpretation between the actual primordial creation (Gen. i. 1), and the chaotic state (ver. 2).

Let us hear an able and eloquent geologist, Professor Sedgwick, on the hypothesis just mentioned of the elongation of the six days :—

" They [certain excellent Christian writers on the subject of Geology] have not denied the facts established by this science, nor have they confounded the nature of physical and moral evidence ; but they have prematurely (and, therefore, without an adequate knowledge of all the facts essential to

* Testimony of the Rocks, p. 144.

the argument) endeavoured to bring the natural history of the earth into a literal accordance with the Book of Genesis; first, by greatly extending the periods of time implied by the six days of creation; and secondly, by endeavouring to show that under this new interpretation of its words, the narrative of Moses may be supposed to comprehend, and to describe in order, the successive epochs of Geology. It is to be feared that truth may, in this way, receive a double injury; and I am certain that the argument just alluded to has been unsuccessful."—" We must consider the old strata of the earth as monuments of a date long anterior to the existence of man, and to the times contemplated in the moral records of his creation."*

Many able theologians, who, though well acquainted with natural science, can scarcely be considered as geologists, have been satisfied with this solution of the problem.

Thus Sharon Turner:—

" What interval occurred between the first creation of the material substance of our globe, and the mandate for light to descend upon it, whether months, years, or ages, is not in the slightest

* Discourse (5th Ed.), 115.

degree noticed [in the Sacred Record]. Geology may shorten or extend its duration, as it may find proper." *

Thus the present Archbishop of Canterbury:—

'" We are not called upon to deny the possible existence of previous worlds, from the wreck of which our globe was organized, and the ruins of which are now furnishing matter for our curiosity." †

Thus Dr. Chalmers:—

" The present economy of terrestrial things was raised about six thousand years ago on the basis of an earth then without form and void; while, for aught of information we have in the Bible, the earth itself may before this time have been the theatre of many lengthened processes, the dwelling-place of older economies that have now gone by, but whereof the vestiges subsist even to the present day, both to the needless alarm of those who befriend Christianity, and the unwarrantable triumph of those who have assailed it." ‡

Thus Dr. Harris:—

" The first verse of Genesis was designed to announce the absolute origination of the material universe by the Almighty Creator; and, passing

* Sac. Hist. of World.   † Rec. of Creation.   ‡ Nat. Theology.

by an indefinite interval, the second verse describes the state of our planet immediately prior to the Adamic creation; and the third verse begins the account of the six days' work." *

Thus Mr. Gray :—

" That an antecedent state of the earth existed before the recorded Mosaical epoch, will clearly come out to view by the consideration of the terms used in the second verse. There was at that period, according to the express Mosaic record, anterior to the six days' reduction into order, *existing earth* and *existing water*." †

Probably the majority of our ablest geologists, men who have devoted their lives to the study and elucidation of geological phenomena, are to be found among those who advocate this scheme of reconciling those phenomena with the statements of the Holy Scriptures. Thus one of the earliest cultivators of the science, the Rev. Dr. Conybeare :—

" I regard Gen. i. 1 as an universal proposition, intended to contradict all the heathen systems which supposed the eternity of matter or polytheism; and ver. 2 I regard as proceeding to take up our planet in a state of ruin from a former condition, and describing a succession of pheno-

* Pre-Adamite Earth.   † Harmony of Scripture and Geology.

mena effected in part by the laws of nature (which are no more than our expression of God's observed method of working), and in part by the immediate exercise of Divine power in directing and creating." *

Dr. Hitchcock, President of Amherst College, U.S., gives in his adhesion to this principle. After summing up the evidence in favour of the earth's high antiquity, he inquires, " Who will hesitate to say that it ought to settle the interpretation of the first verse of Genesis, in favour of that meaning which allows an intervening period between the creation of matter and the creation of light? This interpretation of Genesis is entirely sufficient to remove all apparent collision between Geology and revelation. It gives the geologist full scope for his largest speculations concerning the age of the world. It permits him to maintain that its first condition was as unlike to the present as possible, and allows him time enough for all the changes of mineral constitution and organic life which its strata reveal. It supposes that all these are passed over in silence by the sacred writers, because irrelevant to the object of revelation; but full of interest and instruction to the men of science who

* Christian Observer, 1834.

should afterwards take pleasure in exploring the works of God.

" It supposes the six days' work of creation to have been confined entirely to the fitting up the world in its present condition, and furnishing it with its present inhabitants. Thus, while it gives the widest scope to the geologist, it does not encroach upon the literalities of the Bible; and hence it is not strange that it should be almost universally adopted by geologists, as well as by many eminent divines." *

Dr. Pye Smith, accepting the immense undefined interval between the event of the first verse, and the condition chronicled in the second, held the somewhat remarkable opinion that the term "earth" in that verse, and throughout the whole description of the six days, is " designed to express the part of our world which God was adapting for the dwelling of man and the animals connected with him." And that portion he conceived to have been " a part of Asia, lying between the Caucasian ridge, the Caspian Sea, and Tartary on the north, the Persian and Indian Seas on the south, and the high mountain ridges which run at considerable distances on the eastern and western flank."

* Religion of Geology, Lect. ii.

The whole of the six days' creation was confined, on this hypothesis, to the re-stocking,
with plants and animals, of this limited region
after an inundation caused by its subsidence. The
flood of Noah was nothing more than a second
overflowing of the same region, by " an elevation
of the bed of the Persian and Indian Seas, or a
subsidence of the inhabited land towards the
south." *

The author of " The Protoplast " has made
the very original suggestion, that the geological
periods may have occurred during the paradisaical
condition of man, which he thinks was of an indefinitely protracted duration, human chronology
commencing at the Fall.

" We have no data in Scripture from which to
gather certain information, and Adam may have
lived unfallen *one day*, or *millions of years*." The
years of the first man's mortal life began to be
reckoned when his immortality ceased. He was
nine hundred and thirty years *old :* † he had been

---

* Scripture and Geology.

† I am not *replying* to any of these conflicting opinions ; else,
with respect to this one, I might consider it sufficient to adduce
the *ipsissima verba* of the inspired text. Not a word is said of
Adam's being "nine hundred and thirty years *old ;*" the plain
statement is as follows :—"And *all the days that Adam lived*
were nine hundred and thirty years." (Gen. v. 5.)

nine hundred and thirty years gradually decaying, slowly dying.

"It may, indeed, be said that no man could have survived those convulsions of nature, of which traces have been discovered in the earth's crust. I would reply to this;—First, that we have no reason to suppose that these changes affected the whole globe *at once ;* they may have been *partial and successive ;* and the world's Eden may have been a spot peculiarly exempted from their influence. Secondly, that Adam's body before the fall was not constituted as ours now are ; it was incorruptible and immortal : physical phenomena could have had no deleterious effect upon him." "Why should we find any difficulty in supposing that the geological changes which appear to have passed upon the globe, *after* its creation, and *before* its curse, were to the first man sources of ever-renewing admiration, delight, and advantage ?

"Inclining to the belief that both the animal fall and the animal curse were considerably antecedent to the sin of Adam, I see no difficulty in the admission, that animal death may also have prevailed prior to that event."*

* "Protoplast," pp. 58, 59 ; p. 325 ; 2d. Ed.

While all those writers whose opinions I have cited, feel it more or less incumbent on them to seek a reconciliation between the words of Inspiration and the phenomena of Geology, there are not a few who decline the task altogether. Some eminent in science seem, by their entire avoidance of the question, to allow judgment to go by default. Others more boldly deny that the two can be accommodated.

Mr. Babbage appears to think the archaic Hebrew so insuperably obscure a language, that no confidence can be put in our constructions of its statements; an opinion which, if true, would make the revelation of God to us, with all its glorious types, and promises, and prophecies, more dubious than the readings of Egyptian papyri, or the decipherment of Assyrian cuneiforms.

On this notion, however, Dr. Pye Smith observes: — "All competent scholars, of whatever opinions and parties they may be in other respects, will agree to reject any imputation of uncertainty with respect to the means of ascertaining the sense of the language."

Others find no difficulty in understanding the Hebrew, but in believing it.

C

Professor Baden Powell sees in the plain, un-
varnished narrative of the Holy Spirit, only myth
and poetry : it " was not intended for an historical
narrative " at all; and he thinks (I hope incor-
rectly), that there is a pretty general agreement
with his views.

" Most rational persons," he says, "now acknowledge the failure of the various attempts to reconcile the difficulty [between Geology and Scripture]
by any kind of verbal interpretation ; they have
learnt to see that the ' six days of thousands of
years' have, after all, no more correspondence
with anything in Geology than with any sane
interpretation of the text. And that the ' immense
period at the beginning,' followed by a recent
literal great catastrophe, and final reconstruction in
a week, is, if possible, more strangely at variance
with science, Scripture, and common sense. Yet
while they [viz. the ' rational persons,'] thus view
the labours of the Bible-geologists as fruitless
attempts, they often do not see—," &c. &c. *

Of course this gives up the authority of Scripture altogether; and, consistently enough, the
author is severe upon the prevalent " indiscrimi-
nate and unthinking Bibliolatry." " If in any

* Unity of Worlds (1856), pp. 488, 493.

instance the letter of the narrative or form of expression may be found *irreconcilably at variance with physical truth*,* we may allow, to those who prefer it, the alternative of understanding them either as religious truths, represented under sensible images, or as descriptions of events according to the preconceptions of the writers, or the traditions of the age."

The author of " Vestiges of the Natural History of Creation " propounds a theory of organic origin much more worthy of God, than that " mean view," which supposes Him " to come in on frequent occasions with new fiats or special interferences." Coolly bowing aside His authority, this writer has hatched a scheme, by which the immediate ancestor of Adam was a Chimpanzee, and his remote ancestor a Maggot !

In reviewing this array of opinions, is there not sufficient ground for regarding with caution the claim to certainty which has been boldly put forth

---

* "A geological truth must command our assent as powerfully as that of the existence of our own minds, or of the Deity himself; and any revelation which stands opposed to such truths *must be false.* The geologist has therefore *nothing to do with revealed religion* in his scientific inquiries."—*Edinb. Review,* xv. 16.

for the conclusions of Geology? It cannot be denied that there is here room for a very considerable amplitude ,of choice among discordant hypotheses. All cannot be true, unless on the principle which was claimed for the Church by the Council of Trent—" *Cum enim ecclesia duarum expositionum ubertate gaudeat, non esse eam ad unius penuriam restrigendam!*" I do not for a moment intend to put all these hypotheses and assumptions on the same level. They vary widely as to their tenableness, and as to their prevalence. But if we leave out of view the fears of those who, from insufficient acquaintance with science, are not competent to adjudicate on its positions, and those who despise or decline Biblical authority altogether on this subject, we have still a somewhat wide range to choose from. Shall we accept the *antediluvian*, or the *diluvian* stratification? the six *ages* or the six *days* of creation? the irruptions of internal fire that occurred chiliads *before Man was made*—those during his protracted *paradisaic state*, or those at the time *of the Flood?*— the extension of the Mosaic record *to universal nature*, or its limitation to a region of *south-western Asia?*

I am not blaming, far less despising, the efforts

that have been made for harmonizing the teachings
of Scripture and science. I heartily sympathise
with them. What else could good men do? They
could not shut their eyes to the facts which Geology
reveals: to have said they were not facts would have
been simply absurd. Granting that the whole truth
was before them—the whole evidence—they could
not arrive at other conclusions than those just
recorded; and, therefore, I do not blame their
discrepancy *inter se*. *The true key has not as yet
been applied to the wards.* Until it be, you may
force the lock, but you cannot open it. Whether
the key offered in the following pages will open
the lock, remains to be seen.

II.

THE WITNESS FOR THE MACRO-CHRONOLOGY.

" You shall well and truly try, and a true deliverance make, ... and a true verdict give, according to the evidence."—(Jury Oath.)

A HIGH Court of Inquiry has been sitting now for a good many years, whose object is to determine a chronological question of much interest. It is no less than the age of the globe on which we live. Counsel have been heard on both sides, and witnesses have been called, and most of the judges have considered that an overwhelming preponderance of testimony is in favour of an immeasurably vast antiquity. A single Witness on the other side, however, has deposed in a contrary sense: and, though he has said but little, some of those who have heard the cause attach such weight to his testimony, that they do not feel satisfied to let it be overborne. Counsel on the former side have, indeed, cross-examined the Witness, and dissected his testimony with much skill, and they contend that what he said has been misunderstood by the minority; and that, as his words may at least bear a sense which would not contradict those of the opposing witness, the clear, copious, and unvarying deposition previously made, ought to command the verdict of the Court.

The minority are silenced, but not satisfied; they know not how to give up the Witness on whose veracity they have been wont to rely; but they are unable to answer the arguments brought against him.

Counsel for the Brachy-chronology speaks. "We respectfully ask the Court for another hearing. Will our learned brother permit his witness briefly to recapitulate his testimony, and we will endeavour to examine it once more; for we think we shall be able to detect some flaw in it?" Rule granted.

WITNESS FOR THE MACRO-CHRONOLOGY.

The following, then, is the substance of what the witness deposes.   He is not a living witness; his testimony, therefore, is not oral, but written— lithographed, in fact.   It consists of a number of documents, which are couched in a language and character not to be understood without some previous study, but yet very capable of translation— very clear and unmistakeable. The following, I say, is a condensed summary of the leading points.

If a curious person had watched the process of making the excavations that were preliminary to the boring of the Thames Tunnel, he would have observed that the labourers exposed successive layers of earth, differing much in colour, consistency, and general character. First, an accumulation of soil, consisting of decayed vegetable and animal matter, mingled with broken pottery, and other rubbish of man's production, was removed; then a layer of sand, gravel, and river mud; then a bed of reddish clay; then a layer of clay, mixed with silt or fine sandy mud; then a thin layer of silt, much filled with shells; then a stratum of stiff blue clay; then a layer of clay of more mottled character, containing a portion of silt, and some shells; then a stratum of very firm clay, so solid that it required to be broken with wedges; then a bed of gravel and sand of a green colour; and finally, a similar layer, but of a coarser texture.

In the course of the hundred feet or so of perpendicular depth thus exposed, he would have seen a succession of layers, apparently deposited upon one another. But as yet he would have formed a very inadequate notion of the stratification of the earth's crust.

With the knowledge thus gained, however, let him now make a little excursion into Hertfordshire; we will suppose at the time when the cuttings for the Great Northern Railway were being made. When he came near Cheshunt, he would see that the London clay, which he found underlying the Thames, crops out, or disappears by the stratum coming obliquely to the surface. He would see, however, another bed of clay—the plastic clay—beneath this, which now forms the superficial stratum, and continues to do so, till he gets beyond Hertford. There this stratum crops out; and the chalk, which for some time he has seen to underlie the plastic clay, now comes to the surface.

Business or pleasure calls him to Bridlington on the Yorkshire coast; and he determines to make a pedestrian tour across the diameter of England to Whitehaven. He soon recognises the chalk, which constitutes the Wolds, and rises to

c 3

about 800 feet above the sea level. Below its escarpment he traces the Kimmeridge clay, the uppermost of a series of strata more than 2,000 feet in thickness, that cönstitute the Oolitic system — including, among others, the coralline oolite, the calcareous grit, the cornbrash, thin, but rich in fossils ; the lower sandstone and coal of the Cleveland hills, the alum shale, the marlstone, and the lower lias shale.

Then comes a stratum of the saliferous system or the new red sandstone, with the rèd marls, perhaps not much short of a thousand feet deep. Below them the observer finds the strata of the magnesian limestone formation, for nearly 400 feet, resting on the great coal formations of vast depth. Of these the coal field of the West Riding is not less than 4,000 feet in depth, and beneath it lie the millstone grit, and the mountain limestone, 2,500 feet more, the latter displayed in noble grandeur on the faces of those wall-like precipices that inclose the romantic dales of the Swale and the Ure, and that subsequently tower in magnificent altitude on the sides of Pennygant and Ingleborough.

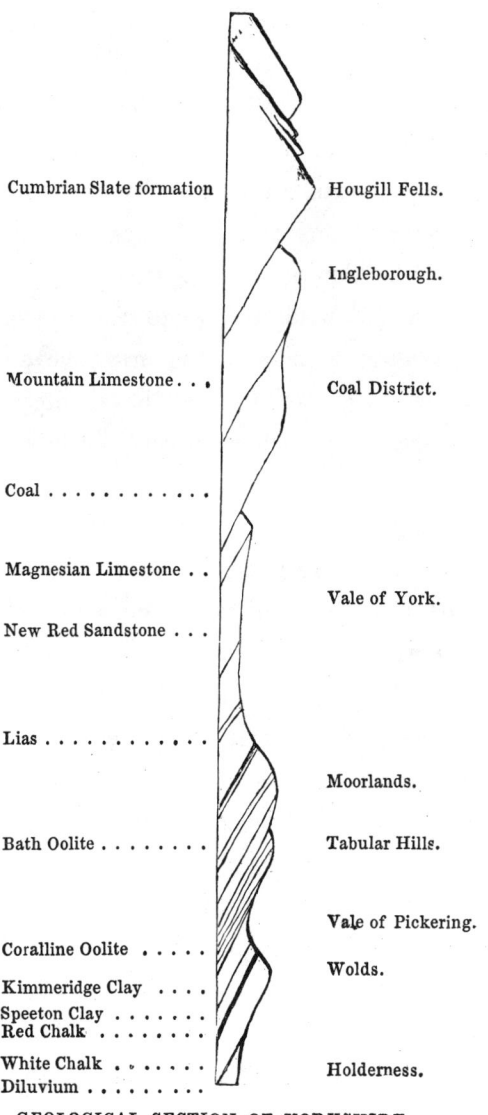

Cumbrian Slate formation | Hougill Fells.

Ingleborough.

Mountain Limestone. . . | Coal District.

Coal . . . .

Magnesian Limestone . . | Vale of York.

New Red Sandstone . . .

Lias . . . . | Moorlands.

Bath Oolite . . . . | Tabular Hills.

Vale of Pickering.

Coralline Oolite . . . . . | Wolds.

Kimmeridge Clay . . . .
Speeton Clay . . . .
Red Chalk . . . .
White Chalk . . . . | Holderness.
Diluvium . . . .

GEOLOGICAL SECTION OF YORKSHIRE.

On the western escarpment of the Pennine ridge, just as the traveller is entering Westmoreland, he would detect the bottom of the limestone; and here he would have an opportunity of seeing, what is rare in these parts, a stratum of the old red sandstone, lying between the former and the slaty rocks of the Cumbrian formations. And here at length, in the wild and magnificent scenery of these mountains, he sees the primitive and transition series, the greenstone, the sienite, and the granite, each of which is discernible in succession on the face of one or other of the lofty Fells of Cumberland.

Our traveller now comes home, and, musing on what he has seen, counts up some thirty or more distinct strata lying in regular succession one on another. But he has not seen all the world, nor even all England; but he reads the results of many independent observations, and finds that while, for the most part, the strata which he has seen are common to the whole surface of the globe, and while the order of their superposition is invariable everywhere, others are in some parts added, while perhaps some of those which he has observed are locally absent. Thus he is able to form a more distinct idea of the stratification of the earth's

crust as a whole. It is composed of about forty
distinct formations, generally increasing in thickness as we go downwards, so that the whole cannot be much less than ten miles in depth, supposing
them in any locality to be all present, and to be
lying in the horizontal plane.

Mathematicians have satisfactorily determined
that the mean density of the globe is about five-
and-a-half times that of water, or about twice that
of granite, a fact inconsistent with any other supposition than that the interior is occupied by
substances maintained in a fluid state by intense
heat. The lowest point that has yet been patent
to human observation is occupied by the granite, a
compound rock, which bears evident marks of
having been once in a state of fusion, and of
having cooled slowly, and that under immense
pressure, contracting and crystallizing as it parted
with its heat. There is every reason to believe
that the granite is not defined at its inferior surface, but that it merges into the molten mass,
probably still solidifying.

After the outer portion of the granite had cooled
sufficiently to become solid, there is evidence that
it was covered by water, agitated by powerful
currents, and probably in a heated state. The

action of these currents disintegrated the rock, and deposited the constituent substances at the bottom of the sea—on the surface, and in the hollows, of the granite. For there is reason to think that the contraction of the primitive rock in the process of cooling, produced irregular undulations or crumplings of the surface, and frequent fractures and dislocations, elevating some parts and depressing others. The gneiss, the mica-schist, and the clay-slate, which are found immediately overlying the granitic rock in strata of vast thickness, are but the components of granite, separated and re-arranged. "If we imagine common granite coarsely pounded, and thrown into a vessel of water, it will arrange itself at the bottom of the vessel in a condition very much like that of gneiss, which is indeed nothing else than stratified granite. If the water in which the pounded rock is thrown is moving along at a slow rate, and the clayey portion of the granite, called *felspar*, happens to be somewhat decomposed, as it often is, then the felspar (which is so truly *clay* that it makes the best possible material for the use of the potteries) and the thin shining plates of mica, will be carried further by the water than the lumps of white quartz or flint sand, which, with the other two ingredients, made up the granite; and the two former will be deposited in layers, which, by passing a galvanic current through them, would in time become mica-schist. If the mica were absent, or if the clay were deposited without it, owing to any cause, then a similar galvanic current would turn the deposit into something like clay-slate."*

The deposition of these strata, being formed out of granite, supposes the pre-existence of that rock; and as they occur in vast thicknesses, even of many thousand feet, then separation, deposition, and re-consolidation must have occupied, however rapidly we may suppose the processes to have been accomplished, considerable periods of time.

In these lower rocks, no trace of organic remains has been found. The shoreless ocean that covered the cooling surface of the earth's crust, harboured no polype or sponge, no rhizopod or infusorium, and the angles and clefts of the granite were fringed by no fucus, or conferva : all was waste and void. And if certain parts were elevated above the waters, the bleak and barren points were not clothed with grass, or moss, or even a lichen, and no animal wandered over their ridges. Or, if such did exist, either in land or water, all vestiges of their presence

* Ansted's Ancient World, 18.

have been destroyed by the agency of the intense heat that subsequently prevailed.

But, in the numerous strata that overlie the rocks of granitic origin, there are found, in varying abundance, proofs that, when they were deposited, the surface of our earth had become the abode of organic life. Zoophytes lived in the ocean, some of which were engaged in secreting lime from the water, and depositing it in coral-reefs; stalked and jointed Star-fishes waved like lilies of stone from the submerged rocks; Sea-worms twined over the mud; mailed Crustaceans swam to and fro; and Mollusks, both bivalve and univalve, crawled over the ledges or reposed in the crevices. The remains of these occur in the Silurian rocks that lie imme- diately on the primitive granitic formations of Cumberland and North Wales. The construction of the coral-reefs of that deposit, in particular, must have occupied a lengthened period, continuing to go on, "month after month, year after year, century after century, until at length the depth changed, in which they could most conveniently live, or, owing to some other cause, their labours were brought to a close, and they disappeared from amongst existing species."*

* Ansted's Ancient World, 30.

Not a single species, or even a single genus of those early strata, is identical with any that exists now. The Coral-polypes, for instance, while allied to ours, are quite distinct from them, though endowed with similar powers and habits, so that we may reason from analogy on the laws of their deposits. The Trilobites were allied to the tiny water-fleas (*Entomostraca*) of the present day : like the *Oniscidæ* (wood-lice, buttons, &c.) of our gardens, they had the habit of rolling their plated bodies into a ball. These are found in great numbers, their remains often heaped on one another.

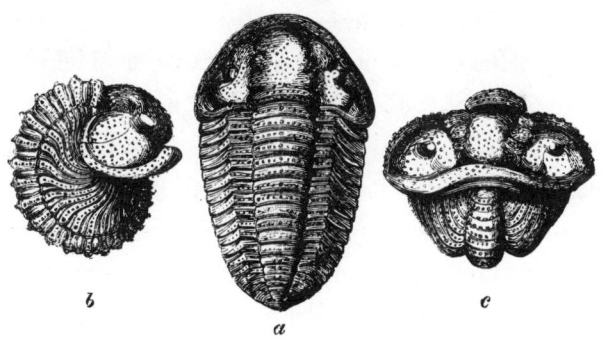

b   *a*   *c*

A TRILOBITE.

(*Calymene Blumenbachii.*)

a. extended; back view.   *b*. rolled up; side view.
c. rolled up ; front view.

The Mollusca of those seas were chiefly of the

class *Cephalopoda*—one of the least populous now-a-days, but then existing in vast number and variety; the Brachiopoda, Conchifera, and Gastropoda,, were, however, well represented also.

Such were the inhabitants of the sea during the Silurian period, in which a series of solid deposits were made, the aggregate, probably, exceeding 50,000 feet in thickness. Each deposit, though not more than a few inches in depth, "is provided with its own written story, its sacred memoranda, assuring us of the regularity and order that prevailed, and of the perfect uniformity of plan."

Over all these, however, we see laid the strata of the Devonian system, especially the old red sandstone, which in some places attains a thickness of 10,000 feet. It is composed of a coarse agglomeration of broken fragments of the old granitic rocks, rolled and tossed about, apparently by the ever-breaking waves of shingle-beaches, until the hardest stones are worn into rounded pebbles by long and constant attrition.

An examination of the old red sandstone, as is seen in Herefordshire, will aid us in forming a notion of the time required for its production. It is composed of fragments obtained by the disintegration of more ancient rocks, which, by a long process of rolling together in a breaking sea, or in the bed of a rapid current, have lost all their angles. The pebbles, thus worn, have at length settled,— the heaviest lowest,—and the whole has been consolidated into firm rock. "In many places," says Dr. Pye Smith, "the upper part of this vast formation is of a closer grain, showing that it was produced by the last and finest deposits of clayey and sandy mud, tinged, as the whole is, with oxides and carbonates of iron, usually red, but often of other hues. But, frequently, the lower portions, sometimes dispersed heaps, and, sometimes, the entire formation, consist of vast masses of conglomerate, the pebbles being composed of quartz, granite, or some other of the earliest kinds; and thus showing the previous rocks, from whose destruction they have been composed. Let any person first acquire a conception of the extent of this formation, and of its depth, often many hundreds, and, sometimes, two or three thousand feet; (but such a conception can scarcely be formed without actual inspection;) then let him attempt to follow out the processes which the clearest evidence of our senses shows to have taken place; and let him be reluctant and sceptical to the utmost that he can, he cannot avoid the impression that ages

innumerable must have rolled over the world, in
the making of this single formation."*

Here, Fishes are added to the Invertebrate Animals. A sort of Shark with the mouth terminal,
instead of beneath the head, was the earliest representative of this class. But closely following on
this, were some curious species, enveloped in plate
mail, and remarkable for the singularity of their
forms, as the *Cephalaspis* and the *Pterichthys*.

CEPHALASPIS.

This great period passed away, and was succeeded
by that of the Carboniferous deposits, indicative
of a vast change in the physical character of the

* Scripture and Geology, 371. (Ed. 1855.)

earth's surface and atmosphere. This change of character may be briefly summed up as consisting of an immense abundance of lime in the ocean, and of an equally vast charge of carbonic acid in the atmosphere.

Strata of limestone, 2,500 feet in thickness, were accumulated in the ocean by the labours of Coral-polypes, allied to, but totally distinct from, those which had previously existed in the primary system. On the floor of a shallow sea, which then occupied the middle of what is now England, the coral reefs rose perpetually towards the day, atom by atom, the strata on which they were founded slowly and steadily sinking ever to a lower level, while successive generations of the industrious zoophytes wrought upwards, to maintain their position within reach of the light and warmth. What period of time was requisite for the aggregation of coral structure to the perpendicular thickness of 2,500 feet?

While this was going on, other Invertebrata were living in the shallow seas, mostly differing from the older species, which had become by this time extinct. Encrinites and Sea-urchins existed; some *Foraminifera* were astonishingly abundant; the *Cephalopoda* and the *Brachiopoda* presented a vast variety of species; and about seventy sorts of Fishes, mostly Sharks, characterised the age.

On the coral limestone lies a sort of conglomerate, known as the millstone grit; and on this is laid that source of Britain's eminence, the *coal.* The coal measures of South Wales are estimated at 12,000 feet in thickness. The profusion of vegetable life that must have combined to make the coal in these, has no parallel in this age; no, not in the teeming forests of South America, or the great isles of the Oriental Archipelago. The circumstances which favoured this enormous development of plants, seem never to have been repeated in subsequent ages, since the coal measures which are found in the later strata are thin and inconsiderable, compared with those we are considering.

M. Adolphe Brogniart suggests that in this period, from some source or other, carbonic acid was generated in vast abundance; or, at least, that it existed in the air, in a far greater proportion than it does now; and it is singularly confirmatory of his view, that terrestrial animals, to which this gas is fatal, have left almost no traces of their existence, during the age of these vast forests—a circumstance otherwise strange and unaccountable.

"Those parts," says Mr. Ansted, "of the great carboniferous series which generally include the beds of coral, consist of muddy and sandy beds, alternating with one another, and with the coal itself. Some of them would appear to be of fresh-water, and some of marine origin; and they abound, for the most part, with remains of the leaves of Ferns and fern-like trees, together with the crushed trunks of these and other trees, whose substance may have contributed to form the great accumulations of bituminised and other vegetable carbon obtained from these strata.

"It is not easy to communicate such an idea of beds of coal as shall enable the reader to understand clearly the nature of the circumstances under which they may have been deposited, and the time required for this purpose. The actual total thickness of the different beds in England varies considerably in different districts, but appears to amount, in the Lancashire coal-field, to as much as 150 feet. In North America there is a coal-field of vast extent, in which there appears at least as great a thickness of workable coal as in any part of England; while in Belgium and France the thickness is often much less considerable, although the beds thicken again still further to the east.

" But this account of the thickness of the beds gives a very imperfect notion of the quantity of vegetable matter required to form them ; and, on the other hand, the rate of increase of vegetables, and the quantity annually brought down by some great rivers, both of the eastern and western continents, is beyond all measure greater than is the case in our drier and colder climate. Certain kinds of trees which contributed largely to the formation of the coal, seem to have been almost entirely succulent, and capable of being squeezed into a small compass during partial decomposition. This squeezing process must have been conducted on a grand scale, both during and after the formation of separate beds ; and each bed in succession was probably soon covered up by muddy and sandy accumulations, now alternating with the coal in the form of shale and grit-stone. Sometimes, trunks of trees caught in the mud would be retained in a slanting or nearly vertical position, while the sands were accumulating round them ; sometimes the whole would be quietly buried, and soon cease to exhibit any external marks of vegetable origin.*

* " It is by no means unlikely that some beds of coal were derived from the mass of vegetable matter present at one time on the surface, and submerged suddenly. It is only necessary

" To relate the various steps in the formation of
a bed of coal, and the gradual superposition of
one bed upon another, by which at length the
whole group of the coal-measures was completed,
would involve an amount of detail little adapted
to these pages ; and when it is remembered that
the woody fibres, after being deposited, had to be
completely changed, and the whole character of
the vegetable modified, before it could be reduced
to the bituminous, brittle, almost crystalline
mineral now dug out of the earth for fuel, it will
rather seem questionable whether the origin of
coal was certainly and necessarily vegetable, than
reasonable to doubt the importance of the change
that has taken place, and the existence of extraordinary means to produce that change. Nothing,
however, is more certain than that all coal was
once vegetable ; for in most cases woody structure
may be detected under the microscope ; and this,
if not in the coal in its ordinary state, at least in
the burnt ashes which remain after it has been
exposed to the action of heat, and has lost its

to refer to the accounts of vegetation in some of the extremely
moist, warm islands in the southern hemisphere, where the
ground is occasionally covered with eight or ten feet of decaying
vegetable matter at one time, to be satisfied that this is at least
possible."

D

bituminous and semi-crystalline character. This has been too well and too frequently proved by actual experiment, to require more than the mere statement of the fact." *

An eminent practical geologist thus essays to guess the age of the coal-fields, and of the sandstone that underlies it.

"The great tract of peat near Stirling has demanded [for its formation] two thousand years; for its registry is preserved by the Roman works below it. It is but a single bed of coal. Shall we multiply it by 100? We shall not exceed,—far from it,—did we allow 200,000 years for the production of the coal-series of Newcastle, with all its rocky strata. A Scottish lake does not shoal at the rate of half a foot in a century; and that country presents a vertical depth of far more than 3,000 feet in the single series of the oldest sandstone. No sound geologist will accuse a computer of exceeding, if he allow 600,000 years for the production of *this series alone.* And yet what are the coal deposits, and what the oldest sandstone, compared to the entire mass of the strata?" †

The conjecture, that the whole of the vegetable

* Ansted's Anc. World, 75.
† M'Culloch's System of Geology, i. 506.

material now constituting the coal, was the growth of the antediluvian centuries, and that it was floated away and deposited by the flood, is untenable. In not a few instances trunks are found broken, and worn by water-action; but the great mass warrants the conclusion that trees of vast dimensions and of close array — dense, majestic forests, such as now occur only in the most humid regions of the tropics—were submerged in their native abodes, lying where they fell, and where they have left the impressions, side by side, on the upper and under surfaces of the shale, of their delicate peculiarities of structure, which would have been totally obliterated, if the trees had been sea-borne and shore-rolled, as pretended. The result of a careful and minute examination of the phenomena of coal, by Mr. Binney, is, that the vegetable matter now forming coal had grown in vast *marine* swamps, subjected to a series of *subsidences* with long intervals of repose ; that the trees, and perhaps smaller plants, were submerged under *tranquil* water, in the places of their growth ; and that very inconsiderable portions, if any, of the beds, are owing to drifting.*

* Origin of Coal.

D 2

While the coal was in process of deposition, the sea was occupied with Invertebrata, not widely differing from those which had marked the previous eras.

Fishes, however, were advancing in development; and several new and strange forms, some of them of gigantic dimensions and formidable armature, were introduced. These were chiefly remarkable for their affinities with Reptiles (whence they are often called *Sauroid* Fishes) ; and one of them—*Megalichthys*—was furnished with jaws of serried teeth, surpassing those of the crocodile. With these were associated other and more ordinary Fishes ; and swarms of Sharks of many species, and varying much in size, roved through the sea, maintaining the same pirate character as their representatives of our modern seas—fierce, subtle, voracious, and powerful.

At this time, too, appeared the earliest Reptiles, chiefly of the Amphibia sub-class. Some of these are known only by their foot-prints ; and the late Hugh Miller has graphically described the appearance of some of these, which he met with marking the roof of a coal-mine, four hundred feet below the surface. These must have been *Batrachia* of large size, as the fore feet were thirteen inches apart across the breast.* They will be alluded to again.

With these exceptions, remains of terrestrial animals are, as has already been observed, rare in this formation.

Testimony of the Rocks, p. 78.

III.

THE WITNESS FOR THE MACRO-CHRONOLOGY.

(CONTINUED.)

" Always distrust very plain cases: beware lest a snake suddenly start out upon you, in the shape of some concealed and utterly unexpected difficulty."
—WARREN: Law Studies.

WE have hitherto been considering the strata as if they had remained permanent when once deposited, subject to no change, save the successive superposition of other strata upon them. But this is very far from being true. Enormous displacements, upheavings, contortions, and fractures, are observed in the strata, which tell of mighty forces having been at work upon them after their formation. The explanation of these phenomena is due to the internal heat, which ever and anon seems to concentrate its action on some special point, seeking and finding vent for itself by some alteration in the already consolidated crust.

Sometimes, the mode of action has been the transmission of undulations through the crust, producing earthquakes, cracking and forcing apart strata already petrified, and bending and variously contorting those that have but partially become solid. Sometimes, the fiery impulse is sufficiently concentrated to break through the superincumbent materials, forcing a passage for the molten and incandescent rock, which then flows forth from the surface, penetrates into the cracks and fissures of the fractured strata, and frequently spreads into the hollows and over the summits of the latest formations.

It is owing to such causes as these, that we find the rocky layers so often inclined at various angles to the horizon, instead of being parallel to it, as they would be of course deposited; occasionally standing quite perpendicularly, and even to a small extent reversed. The outcropping of formations, the long lines of cliff running across a country in parallel series, ("crag and tail,") the dipping of strata from some central point or ridge, and the non-correspondence between the bottom of one stratum and the top of the underlying one,—are all phenomena of this sort of powerful action, which has been more or less energetic at all periods.

After the deposit of the Old Red Sandstone, the internal fire appears to-have enjoyed a lull of its energy, if not a complete cessation, until the Coal Measures were complete. Then the long tranquility was again broken, and concussions so extensive and violent ensued, that hardly a single square mile of country can anywhere be found which is not full of fractured and contorted strata, the record of subterranean movements, which mostly occurred between the Carboniferous and the Premian deposits.

The effects of these convulsions were manifest in the changed relations of land and sea, existing continents and islands being dislocated, severed, and swallowed up, while others were elevated from the depths of the previous ocean.

It was from the wave-worn materials thus obtained from pre-existing strata, that the New Red Sandstone was consolidated. It consists chiefly of sand and mud, with few organic remains; and the hiatus thus found, in animals and vegetables, seems to be almost a complete one between the organisms of the preceding and the succeeding periods.

The most interesting traces of the earth's tenants during the New Red formation, consist of foot-tracks impressed by the progress of animals along the yielding mud between the ranges of high and low tide. They afford a remarkable example (not, I think, sufficiently dwelt on) of the extreme rapidity with which deposits were consolidated; since the tracks must have been made, and the material consolidated, during the few hours, *at most*, that intervened between the recess and the reflux of the tide; since, if the mud had not so soon become solid, the flow of the sea would have instantly obliterated such marks, as it does now on our shores.

LABYRINTHODON PACHYGNATHUS.

The principal animal, whose footprints have been identified, was an enormous Frog (*Labyrinthodon*), as big as a hippopotamus, but apparently allied, in its serried teeth, and in the bony plates with which

D 3

it was covered, to the Crocodiles, which were its associates.

It is curious that marks in the same material have chronicled the serpentine trail of a Sea-worm, the scratchings of a Crab, the ripple of the wavelets, and even the drops of a passing shower; the last revealing, by their margins, the direction of the wind by which the slanting rain was driven.

If the Triassic formations display but little evidence of organic existence, the lack is supplied by the abundance of such records, which is contained in the Oolitic system, and specially in its lowest component,—the Lias. Animals now existed in profusion, but of species which were for the most part peculiar. The coral-making Polypes existed not (or very rarely) in the seas of that age, but lime was secreted by an unusual number of Crinoid Echinoderms, which seem to have fringed the rocks and floating pieces of timber, much as Barnacles do now.

Among the Mollusca now began to appear the inhabitants of those very elegant shells, the *Ammonites,* allied to the Nautilus of our Southern seas, which may be considered as the lingering representative of those swarms of shelled Cephalopoda. They were accompanied by their near relations, the *Belemnites*, more resembling a Cuttle, with a long internal, pointed shell.

Fishes, chiefly belonging to a curiously armed tribe of Sharks, together with some enclosed in bony-mail like pavement, were present in the shallows, where the Lias was probably deposited.

But the most characteristic animals were great marine Reptiles, of strange and uncouth forms, to

SNAKE-NECKED MARINE LIZARDS.
Plesiosaurus dolichodeirus and *P. macroeephalus.*

which the present world presents us no known analogy. One of these was the *Ichthyosaurus*, which closely resembled a porpoise in form, but thirty or forty feet in length, with a vertical fish-like tail, and two pairs of paddles; a mouth set with stout crocodilian teeth, and enormous eyes. Another form was that of the *Plesiosaurus*, scarcely less in size than its fellow, which in the outline of its body it resembled: it was distinguished, however, by an extraordinary length of neck, slender and swan-like, consisting of thirty or forty vertebræ.

It adds to the interest of these great marine Reptiles, that around their fossil skeletons are preserved pellets of excrement (known as Copro-lites) containing fragments of bone, teeth, and scales of fishes, which clearly reveal the nature of their food. In some instances, the stomach and intestines of these great carnivorous creatures, filled with half-digested food, have left indubitable traces of their presence *in situ*.

Again, the geography of the Globe changed. New lands arose from the sea, and old lands partially or wholly sank. The German Ocean, and part of Western Europe, of our maps, were a great group of islands. The Oolitic formation was deposited. The general character of the organization of this period differed little from that of the Lias. New forms of plants, such as *Cycadeæ*, were abundant, with considerable numbers of Corals,

Encrinites, Sea-urchins and Mollusks. Macrurous Crustacea, much like those of our times (but essentially different in species), inhabited the sea, and some Beetles and Flies represented the Insects of the land. The Fishes and Marine Reptiles were pretty much the same with those of the Lias, though they received some important additions.

MEGALOSAURUS BUCKLANDI.

It is, however, among the terrestrial Vertebrata that we must look for the characteristic organisms of this age. And these are, still, Reptiles. The huge *Megalosaurus*, with a body as big as an elephant's, stood high on his legs, and stretched open a pair of gaping jaws, set with jagged teeth.

The *Pterodactyles* flew about,—carnivorous lizards, with the body and wings of bats,* except that the

BAT-LIZARDS.
Pterodactylus crassirostris, and *P brevirostris.*

membrane was stretched upon the enormously developed little finger;—creatures, perhaps, the

* Mr. Newman suggests that they were "marsupial bats" (Zoologist, p. 129). I have adopted his attitudes, but have not entured to give them mammalian ears.

most unlike to anything familar to us, of all fossil
forms. And, in the marshy margins of the great
river valley which formed the Wealden of our South-
eastern districts, the giant *Iguanodon*, and his fellow,
the *Hylæosaurus*, waged their peaceful warfare on
the succulent plants that became their unresisting
prey.

HYLÆOSAURUS ARMATUS.

The circle of animal life was completed in this
epoch, thus far, that every class was represented by
some one or more of its constituent species. No
fossil skeletons of Birds have, indeed, been found
so low as the Oolite, but numerous foot-prints of
some of the Grallatores are found in a sandstone
of this period; and in the Stonesfield slate, which

is contemporary with it, a genus of Mammalia has been discovered,—a small Marsupial, allied to the Opossums of America.

The duration of the Oolitic period must have been considerable. " The lias sea-bottom was succeeded first by a sandy, and then by a calcareous deposit, and the animals were modified accordingly." The deposit of carbonate of lime, which took place under circumstances that caused it to attract around its nodules the organic particles, whence the name *oolite* (egg-stone) is derived, was not continuous, but repeated at intervals. The shells of Mollusks were developed in great abundance, and accumulations of these formed thick bands, which consolidated into layers of shell-limestone. Three hundred feet of strata, largely composed of organic remains, were formed before the clay was deposited which made the Stonesfield and contemporaneous slates.

Once more the dry land sank, probably by slow successive subsidences, and the sea flowed many fathoms deep above the great European archipelago. And upon its quiet bottom settled down, first a few sandy and clayey beds, and then the great layer of the Chalk.

Creatures of very minute size and low grades of

organization were now playing a very important part. A large portion of the lime that was deposited, in the form of a pure carbonate, was doubtless supplied by the Coral structures, which were exceedingly numerous; the polypidoms being gnawed down by strong-jawed fishes that fed upon the Zoophytes. *Foraminifera* also were abundant, and contributed to the supply.

Nodules of flint exist in the Chalk, sometimes scattered, sometimes arranged in bands. Two sources are indicated for this substance. One is Sponge, the most common kinds of which are composed of skeletons of siliceous spicula; and these can be discerned with the microscope in the interior of the chalk-flints. But millions upon millions of Infusoria swam through the waters, and many of these were encased in siliceous loricæ, while the rocks and seaweeds were fringed with as incalculably numerous examples of siliceous *Diatomaceæ*, whose elegant forms are recognisable without difficulty throughout the Chalk. The inconceivable abundance of these forms may be illustrated by the often-cited fact, that whole strata of solid rock appear to be so exclusively composed of their solid remains, that a cube of one-tenth of an inch is computed by Ehrenberg to contain five hundred millions of individuals.

The increase of these organisms is very rapid, and their duration proportionately short; but allowing for this, what period would elapse before the successive generations of entities, of which forty-one thousand millions are required to make a cubic inch, would have accumulated into solid strata fourteen feet in thickness?

Without pausing to examine the whole Cretaceous fauna, we may observe that the Mollusca with chambered shells—the Ammonites and their allies—were developed in singular variety and profusion during this period, after which they suddenly disappeared from the ocean. The Fishes present little that is remarkable; of Birds, few, and of Mammals, no remains exist; and the Reptiles, while not absolutely extinct, are few and rare. One great marine form, however, the *Mosasaurus*, was added to their number.

At length the sea ceased to deposit chalk, and its bed appears to have been slowly elevated, until all the animals that had inhabited the waters of that formation were destroyed; so that their race and generation perished.* The grand epoch of Secondary Formations was closed.

* In Tennant's "List of Brit. Fossils" (1847), but two species —a Brachiopod and a Gastropod—are mentioned as common to the Chalk and the London Clay. They are *Terebratula striatula*, and *Pyrula Smithii*.

It was followed by an extensive disruption of the then existing strata, and by changes and modifications so great as to alter the whole face of nature. " It would appear that a long period of time elapsed before newer beds were thrown down, since the chalky mud not only had time to harden into chalk, but the surface of the chalk itself was much rubbed and worn." During this protracted period, eruptions of molten rock occurred of enormous extent, producing the Basaltic formation which covers the Chalk in the north of Ireland, and in some of the Hebrides. In the south of Europe the Pyrenees were elevated, and the Apennines and Carpathians were pushed to a greater altitude than before, if they were not then formed. The Alps and the Caucasus also experienced a series of upward movements, continuing through a considerable range of the Tertiary epoch.

The rich collections of vegetable remains—chiefly fruits and seeds—that have been made from the London Clay, show that the earliest land of this period was clothed with a great abundance and variety of plants; and these are of such alliances as would now require a tropical climate. Many species of Palms, Screw-pines, Gourds, *Piperaceæ*, *Mimoseæ*, and other *Leguminosæ*, *Malvaceæ*, and

Coniferæ, dropped their woody pods and fruits where now these pages are written; and the animals manifest no less interesting an approximation to existing forms than the plants. The Zoophytes, the Echinoderms, the Foraminifera, the Worms, the Crustacea, the Mollusca, the Fishes and the Reptiles of the Eocene beds, exhibit a great preponderance of agreement with those that now exist, *so far as genus is concerned*, though the *species* are still almost wholly distinct. The approximation is particularly marked in the Molluscous sub-kingdom, by the almost entire disappearance of the hitherto swarming Brachiopod and Cephalopod forms, and the progressive substitution for them of the *Conchifera* and *Gastropoda*, which had, however, throughout the Secondary epoch, been gradually coming forward to their present predominance in nature.

Among the Fishes, the Placoid type was diminished in number; and those that were produced were mostly Sharks and Rays, of modern genera; but the chief difference was the paucity of those mailed forms (Ganoids), which were so abundant during the Oolitic period. On the other hand, the Ctenoid and Cycloid forms, which had begun to make their appearance in small numbers in the

Chalk, are well represented. In both this deficiency and this plenitude, there is a very decided approach to existing conditions; for the Ganoids are almost unknown with us, while the last-named two orders are abundant. Representatives of our Perches, Maigres, Mackerels, Blennies, Herrings, and Cods, were numerous; *distinct, however, from the present species*. But not a single member of the great Salmon family was yet introduced.

The great Saurian Reptiles had entirely disappeared, and were quite unrepresented in the tertiary beds, except by a Crocodile or two, and a small Lizard. Turtles were, however, numerous, both of the marine and lacustrine kinds; and there is an interesting stranger, in the form of a large Serpent, allied to our Pythons, some twenty feet in length.

Birds and Mammals began now to assume their place on the land. The London Clay presents us with a little Vulture; and the Paris basin contains remnants of species representing the Raptores, the Rasores, the Grallatores, and the Natatores.

The Quadrupeds came in in some force; not developed from the lowest to the highest scale of organization; for the Monkey and the Bat occur in sands, certainly not later, if not earlier, than the

London Clay, contemporaneously with the Racoon, and before the existence of any Rodent or Cetacean. Some Carnivora, as the Wolf and the Fox, roamed the woods, but the character of the epoch was given by the Pachyderms.

These, however, were not the massive colossi that browse in the African or Indian jungles of our days; no Elephant, no Rhinoceros, no Hippopotamus was as yet formed. But several kinds of Tapir wallowed in the morasses; and a goodly number of largish beasts, whose affinities were with the Pachydermata, while their analogies were with the Ruminantia, served as substitutes for the latter order, which was wholly wanting. These interesting quadrupeds, forming the genus *Anoplotherium*, were remarkable for two peculiarities,—their feet were two-toed, and their teeth were ranged in a continuous series, without any interval between the incisors and the molars. They varied in size from that of an ass to that of a hare.

The physical conditions of our earth, when it was tenanted by these creatures, is thus described: —" All the great plains of Europe, and the districts through which the principal rivers now run, were then submerged; in all probability, the land chiefly extended in a westerly direction, far out into the

Atlantic, possibly trending to the south, and connecting the western shores of England with the volcanic islands off the west coast of Africa. The great mountain chains of Europe, the Pyrenees, the Alps, the Apennines, the mountains of Greece, the mountains of Bohemia, and the Carpathians, existed then only as chains of islands in an open sea. Elevatory movements, having an east and west direction, had, however, already commenced, and were producing important results, laying bare the Wealden district in the south-east of England. The southern and central European district, and parts of western Asia, were the recipients of calcareous deposits (chiefly the skeletons of *Foraminifera*), forming the Apennine limestone; while numerous islands were gradually lifted above the sea, and fragments of disturbed and fractured rock were washed upon the neighbouring shallows or coast-lines, forming beds of gravel covering the Chalk. The beds of Nummulites and Miliolites, contemporaneous with those containing the Sheppey plants and the Paris quadrupeds, seem to indicate a deep sea at no great distance from shore, and render it probable that there were frequent alternations of elevation and depression, perhaps the result of disturbances acting in the direction already alluded to.

" The shores of the islands and main land were, however, occasionally low and swampy, rivers bringing down mud in what is now the south-east of England, and the neighbourhood of Brussels, but depositing extensive calcareous beds near Paris. Deep inlets of the sea, estuaries, and the shifting mouths of a river, were also affected by numerous alterations of level not sufficient to destroy, but powerful enough to modify, the animal and vegetable species then existing; and these movements were continued for a long time." *

After the elevation of the mountain summits of Europe above the sea, and while the same causes were still in operation, deposits were being made in the narrow intervening seas of the Archipelago, such as the present south of France, the valleys of the Rhine and Danube, the eastern districts of England and Portugal. These deposits were partly marine and partly lacustrine; the former consisting largely of loose sands, mingled with shells and gravel. In Switzerland is a thick mass of conglomerate; and in the district around Mayence, there is a series of fresh-water lime-stones, and sandstones charged with organic remains.

The changes which took place during this com-

* Ansted's Anc. World, 267.

paratively recent epoch were not sudden, but gradual; the results of operations which were probably going on without intermission, and perhaps have not yet ceased. The land was more and more upheaved, till at length, what had been an archipelago of islands became a continent, and Europe assumed the form which it bears on our maps.

The most interesting addition to the natural history of the Miocene, or Middle Tertiary period, was the *Dinotherium*—a huge Pachyderm, twice as large as an elephant, with a tapir-like proboscis, and two great tusks curving downward from the lower jaw. It was, doubtless, aquatic in its habits, and possibly (for its hinder parts are not known), it may have been allied to the Dugong and Manatee, those whale-like Pachyderms, with a broad horizontal tail, instead of posterior limbs.

Other great herbivorous beasts roamed over the new-made land. The Mastodons, closely allied to the Elephant, had their head-quarters in North America, but extended also to Europe. And the Elephants themselves, of several species, were spread over the northern hemisphere, even to the polar regions. The Hippopotamus, the Rhinoceros, and other creatures, now exclusively tropical, were also inhabitants of the same northern latitudes.

MAMMOTH.

From some specimens of Elephants and Rhinoceroses of this period, which seem to have been buried in avalanches, and thus to have been preserved from decomposition, even of the more transitory parts, as muscle and skin, we learn something of the climate that prevailed. The very fact of their preservation, by the antiseptic power of frost, shows that it was not a tropical climate in which they lived; and the clothing of thick

wool, fur, and hair, which protected the skin of
the Mammoth, or Siberian Elephant, tends to the
same conclusion. At the same time, those regions
were not so intensely cold as they are now. For
the district in which the remains of Elephants
and their associates are found, in almost incredible
abundance, is that inhospitable coast of northern
Asia which bounds the Polar Sea.

The trees of a temperate climate—the oak, the
beech, the maple, the poplar, and the birch—which
now attain their highest limit somewhere about 70°
of north latitude, and there are dwarfed to minute
shrubs, appear then to have grown at the very
verge of the polar basin; and that in the condition
of vast and luxuriant forests, perhaps occupying
sheltered valleys between mountains whose steep
sides were covered with snow, already become
perennial, and ever and anon rolling down in
overwhelming avalanches, such as those which
now occasionally descend into the valleys of the
Swiss Alps.

The coast of Suffolk displays a formation known
as the Crag—a local name for gravel—which rests
partly on the chalk; but, as it lies in other parts
over the London Clay, it is assigned to the later
Tertiary, or what is called the Pleiocene period. It

is divided into the *coralline* and the *red* crag, the latter being uppermost where they exist together, and therefore being the more recent. The Coralline Crag is nearly composed of corals and shells, the former almost wholly extinct now; but the latter containing upwards of seventy species still existing in the adjacent seas. The Red Crag contains few zoophytes, but is remarkable for the remains of at least five species of Whales. Other Mammalia occur in this formation, among which are the red deer and the wild boar of modern Europe.

The gradual but rapid approximation of the Tertiary fauna to that of the present surface is well indicated by Mr. Lyell's table (1841) of recent and fossil species in the English formations:—

Periods.	Localities.	Per-centage of recent.	No. of fossils compared.
Eocene	London and Hampshire	1 or 2	400
Miocene . . .	Red and Coralline Crag, Suffolk . . .	20 to 30	450
Older Pleiocene .	Mamaliferous or Norwich Crag	60 to 70	111
Newer Pleiocene.	Marine strata near Glasgow	85 to 90	160
Post Pleiocene .	Fresh-water of the valley of the Thames .	99 to 100	40

It is to this period that are assigned the animals whose bones are found in astonishing numbers in limestone caverns, as, for example, that notable one at Kirkdale, in Yorkshire, which was examined by Professor Buckland.

This is a cave in the Oolitic limestone, with a nearly level floor, which was covered with a deposit of mud, on which an irregular layer of sparry stalagmite had formed by the dripping of water from the low roof, carrying lime in solution. Beneath this crust the remains were found.

Of the animals to which the bones belonged, six were *Carnivora*, viz. *hyæna*, *felis*, bear, wolf, fox, weasel; four *Pachydermata*, viz. elephant, rhinoceros, hippopotamus, horse; four *Ruminantia*, viz. ox, and three species of deer; four *Rodentia*, viz. hare, rabbit, water-rat, mouse; five Birds, viz. raven, pigeon, lark, duck, snipe.

The bones were almost universally broken; the fragments exhibited no marks of rolling in the water, but a few were corroded; some were worn and polished on the convex surface; many indented, as by the canine teeth of carnivorous animals. In the cave the peculiar excrement of hyænas (*album græcum*) was common; the remains of these predacious beasts were the most abundant of all the bones; their teeth were found in every condition, from the milk-tooth to the old worn stump; and from the whole evidence Dr. Buckland adopted the conclusion, in which almost every subsequent writer has acquiesced, that Kirkdale Cave was a den of hyænas during the period when elephants and hippopotami (not of existing species) lived in the northern regions of the globe, and that they dragged into it for food the bodies of animals which frequented the vicinity.*

Thus in these spots we find, observes Professor Ansted, " written in no obscure language, a portion of the early history of our island after it had acquired its present form, while it was clothed with vegetation, and when its plains and forests were peopled by many of the species which still exist there ; but when there also dwelt upon it large carnivorous animals, prowling about the forests by night, and retiring by day to these natural dens."

In our own country, and in many other parts of the world, we find fragments of stone distributed over the surface, sometimes in the form of enormous blocks, bearing in their fresh angles evidence that they have been little disturbed since their disruption, but sometimes much rubbed and worn, and broken into smaller pieces, till they form what is known as gravel. In many cases the original

* Reliquiæ Diluvianæ.

rock from which these masses have been separated
does not exist in the vicinity of their locality; and
it is not till we reach a distance, often of hundreds
of miles, that we find the formation of which they
are a component part.

Various causes have been suggested for the
transport of these erratic blocks, of which the most
satisfactory is the agency of ice, either as slow-
moving glaciers, or as oceanic icebergs.

" The common form of a glacier," says Professor
J. Forbes, " is a river of ice filling a valley, and
pouring down its mass into other valleys yet
lower. It is not a frozen ocean, but a frozen
torrent. Its origin or fountain is in the ramifications of the higher valleys and gorges, which
descend amongst the mountains perpetually snow-
clad. But what gives to a glacier its most peculiar
and characteristic feature is, that it does not belong
exclusively or necessarily to the snowy region
already mentioned. The snow disappears from its
surface in summer as regularly as from that of the
rocks which sustain its mass. It is the prolongation or outlet of the winter-world above; its gelid
mass is protruded into the midst of warm and pine-
clad slopes and green-sward, and sometimes reaches
even to the borders of cultivation." *

* Travels through the Alps, p. 19.

The glacier moves onward with a slow but steady march towards the mouth of its valley. Its lowest stratum carries with it numerous fragments of rock, which, pressed by the weight of the mighty mass, scratch and indent the surfaces over which they move, and sometimes polish them. These marks are seen on many rock-surfaces now exposed, and they are difficult to explain on any other hypothesis than that of glacial action.

But the alternate influence of summer and winter—the percolation of rain into the mountain fissures, and the expansion of freezing—dislodge great angular fragments of rock, which fall on the glacier beneath. Slowly but surely these then ride away towards the mouth of the valley, till they reach a point where the warmth of the climate does not permit the ice to proceed; the blocks then are deposited as the mass melts. But if the climate itself were elevated, or if the surface were lowered so as to immerse the glacier in the sea, it would melt throughout its course, and then the blocks would be found arranged in long lines or *moraines*, such as we see now in many places.

If the glacier-valley debouch on the sea, the ice gradually projects more and more, until the motions of the waves break off a great mass, which floats away, carrying on its surface the accumulation of boulders, gravel, and other *débris* which it had acquired during its formation. It is now an iceberg, which, carried by the southern currents, approaches a warmer climate, melts, and deposits its cargo, perhaps hundreds of leagues from the valley where it was shipped, and as fresh as when its component *frusta* were detached from the primitive rock.

If the abundance of such erratic blocks and foreign gravel seem to require a greater amount of glacial action than is now extant, it has been suggested that the volcanic energy which elevated Europe may have been succeeded by a measure of subsidence before the land attained its present permanent condition. Hence there may have been, during the Tertiary epoch, mountain chains of great elevation, sufficient to supply the glaciers, which, on their subsidence, melted on the spot where they were submerged, or floated away as icebergs on the pelagic currents, till they grounded on the bays and inlets of other shores, which were subsequently elevated again.

Thus a large portion of the animals which then inhabited these islands (up to that time, perhaps, united to the continent) would be drowned, and many species quite obliterated, a few alone remaining to connect our present fauna with that of the submerged area, when the land rose again to its existent state.

It would not materially augment the force of the evidence already adduced on the question of chronology, to examine in detail the fossil remains of South America, Australia, and New Zealand. The gigantic Sloths * of the first, the gigantic Marsupials of the second, and the gigantic Birds of the third, however interesting individually, and especially as showing that a prevailing type governed the fauna in each locality then as now— are all formations of the Tertiary period, and some of them, at least, seem to have run on even into the present epoch.   Indeed, it is not quite certain that the enormous birds of New Zealand and Madagascar are even yet extinct.

* Prof. Owen, in his admirable account of the *Mylodon,* has mentioned a fact which brings us very vividly into contact with its personal history. He shows that the animal got its living by overturning vast trees, doing the work by main strength, and feeding on the leaves. The fall of the tree might occasionally put the animal in peril; and in the specimen examined there is proof of such danger having been incurred. The skull had undergone two fractures during the life of the animal, one of which was entirely healed, and the other partially. The former exhibits the outer tables of bone broken by a fracture four inches long, near the orbit. The other is more extensive, and behind, being five inches long, and three broad, and over the brain. The inner plate had in both these cases defended the brain from any serious injury, and the animal seems to have been recovering from the latter accident at the time of its death.

The phenomenon of raised sea-beaches is one of great interest, and seems to be connected with the alternate elevations and depressions of the Tertiary epoch, perhaps marking the successive steps of the upheaval of the land. In several parts of England the coast-line exhibits one or more shelves parallel with the existing sea-beach, and covered with similar shingle, sand, and sea-shells. And the same phenomenon is exhibited on a still more gigantic scale in South America. Mr. Darwin * found that for a distance of at least 1,200 miles from the Rio de la Plata to the Straits of Magellan on the eastern side, and for a still longer distance on the west, the coast-line and the interior have been raised to a height of not less than 100 feet in the northern part, but as much as 400 feet in Patagonia. All this change has taken place within a comparatively short period; for in Valparaiso, where the effect is most considerable, modern marine deposits, with human remains, are seen at the height of 1,300 feet above the sea.

At what exact point, geologically, the period of human history begins, it is impossible to say. No evidence of Man's presence has occurred older than the latest Tertiary deposits, which insensibly merge

* Naturalist's Voyage, *passim.*

into the Alluvial. It seems certain that human remains have been found in chronological association with those of animals long extinct, and there appears no reason to doubt that some species of animals, as the Irish Deer, the Moa of New

THE MOHO.

Zealand, and the Dodo of the Mauritius, have disappeared from creation within a period of a few centuries.* It is not improbable that the last of

---

* The Indians of North America knew that the Mastodon had a trunk ; a fact which (though the anatomist infers it from

the Moho race may have lived only long enough
to grace the pages of the " Birds of Australia."

It is as important as it is interesting, to observe
that the same kinds of physical operations have
been, within the present epoch, and are still, going
on, as those whose results are chronicled in the
rocks. Strata of alluvium are constantly being
formed on a scale which, though it does not
suddenly affect the outline of coasts, and therefore
appears small, yet is great in reality.

The Ganges is estimated to pour into the Indian
Ocean nearly 6,400 millions of tons of mud every
year; and its delta is a triangle whose sides are
two hundred miles long. The delta of the Mississippi is of about the same size, and it advances
steadily into the Gulf of Mexico at the rate of
a mile in a century.

The accumulation of river-mud is gradually
filling up the Adriatic Sea. From the northern-
most point of the Gulf of Trieste to the south of
Ravenna, there is an uninterrupted series of recent
accessions of land, more than a hundred miles in
length, which, within the last twenty centuries,

the bones of the skull) it is difficult to imagine them to be
acquainted with, except by tradition from those who had seen
the living animal.

have increased from two to twenty miles in breadth.

The coral-polypes are working still with great energy. Mr. Darwin mentions two or three examples of the rate of increase, one of which only I shall cite. In the lagoon of Keeling Atoll, a channel was dug for the passage of a schooner built upon the island, through the reef into the sea; in ten years afterward, when it was examined, it was found almost choked up with living coral.

Volcanic action is busy in many parts of the earth, pouring forth clouds of ashes and torrents of molten rock; and instances are not wanting in which new islands have been raised from the bed of the ocean by this means, within the sphere of history.

Slow and permanent changes of level are still being produced on the earth's crust. The bottom of the Baltic has been, for several centuries at least, in process of continuous elevation, the effects of which are palpable. Many rocks formerly covered are now permanently exposed; channels between islets, formerly used, are now closed up, and beds of marine shells have become bare. On the other hand, the whole area of the Pacific Polynesia seems subsiding.

Deposits are being made by waters which hold earthy substances in solution. The principal of these is *lime*. Several remarkable examples of this kind are quoted by Sir Charles Lyell, in one of which there is a thickness of 200 or 300 feet of travertine of recent deposit, while in another a solid mass thirty feet thick was deposited in about twenty years. He also states that there are other countless places in Italy where the constant formation of limestone may be seen, while the same may be said of Auvergne and other volcanic districts. In the Azores, Iceland, and elsewhere, *silica* is deposited often to a considerable extent. Deposits of *asphalt* and other bituminous products occur in other places.*

The floors of limestone caverns are frequently strewn with fossil bones, which are imbedded in stalagmite, and this incrustation is still in progress of formation. It is remarkable that in this deposit alone we obtain the bones of Man in a fossil condition. The two creations,—the extinct and the extant,—or rather the prochronic and the diachronic—here unite. But there is no line of demarcation between them; they merge insensibly into each other. The bones of Man, and even his

* Ansted; Phys. Geography, 82.

implements and fragments of pottery, are found mingled with the skeletons of extinct animals in the caves of Devonshire, in those of Brazil,* and

* An interesting fact relating to the Brazilian caves was communicated to Dr. Mantell. "M. Claussen, in the course of his researches, discovered a cavern, the stalagmite floor of which was entire. On penetrating the sparry crust, he found the usual ossiferous bed; but pressing engagements compelled him to leave the deposit unexplored. After an interval of some years, M. Claussen again visited the cavern, and found the excavation he had made completely filled up with stalagmite, the floor being as entire as on his first entrance. On breaking through this newly-formed incrustation, it was found to be distinctly marked with lines of dark-coloured sediment, alternating with the crystalline stalactite. Reasoning on the probable cause of this appearance, M. Claussen sagaciously concluded that it arose from the alternation of the wet and dry seasons. During the drought of summer, the sand and dust of the parched land were wafted into the caves and fissures, and this earthy layer was covered during the rainy season by stalagmite, from the water that percolated through the limestone, and deposited calc-spar on the floor. The number of alternate layers of spar and sediment tallied with the years that had elapsed since his first visit; and on breaking up the ancient bed of stalagmite, he found the same natural register of the annual variations of the seasons; every layer dug through presented a uniform alternation of sediment and spar; and as the botanist ascertains the age of an ancient dicotyledonous tree from the annual circles of growth, in like manner the geologist attempted to calculate the period that had elapsed since the commencement of these ossiferous deposits of the cave; and although the inference, from want of time and means to conduct the inquiry with precision, can only be accepted as a rough calculation, yet it is interesting to learn that the time indicated by this natural chronometer, since the extinct mammalian forms were interred, amounted to many thousand years."—(*Petrifactions and their Teachings*, p. 481.)

in those of Franconia. In Peru, some scores of human skeletons have been found in a bed of travertine, associated with marine shells; the stratum itself being covered by a deep layer of vegetable soil, forming the face of a hill crowned with large trees.

From a very interesting paper by M. Marcel de Serres, it appears indubitable that the existing shells of the Mediterranean are even now passing in numbers into the fossil state, and that not in quiet spots only, but where the sea is subject to violent agitations. Specimens of common species, "completely petrified, have been converted into carbonate of lime at the same time that they have lost the animal matter which they originally contained. Their hardness and solidity are greater than those of some petrified species from tertiary formations."

" In the collection of M. Doumet, Mayor of Cette, there exists an anchor which exhibits the same circumstances, and which is also covered with a layer of solid calcareous matter. This contains specimens of *Pecten*, *Cardium*, and *Ostrea*, completely petrified, and the hardness of which is equal to that of fossil species from secondary formations. On the surface of the deposit in which the anchor is imbedded, there are *Anomiæ* and *Serpulæ*, which were living when the anchor was got out of the sea; these present no trace of alteration." *

Thus we have brought down the record to an era embraced by human history, and even to individual experience; and we confidently ask, Is it possible, is it imaginable, that the whole of the phenomena which occur below the diluvial deposits can have been produced within six days, or seventeen centuries? Let us recapitulate the principal facts.

1. The crust of the earth is composed of many layers, placed one on another in regular order. All of these are solid, and most are of great density and hardness. Most of them are of vast thickness, the aggregate not being less than from seven to ten miles.

2. The earlier of these were made and consolidated before the newer were formed; for in several cases, it is demonstrable that the latter were made out of the *débris* of the former. Thus the compact and hard granite was disintegrated grain by grain; the component granules were rolled awhile in the sea till their angles were rubbed down; they were slowly deposited, and then consolidated in layers.

* Bibliothèque Univers., March, 1852.

3. A similar process goes on again and again to form other strata, all occupying long time, and all presupposing the earlier ones.*

4. After some strata have been formed and solidified, convulsions force them upward, contort them, break them, split them asunder. Melted matter is driven through the outlets, fills the veins, spreads over the surface, settles into the hollows, cools and solidifies.

5. After the outflowing and consolidation of these volcanic streams, the action of running water cuts them down, cleaving beds of immense depth through their substance. Mr. Poulett Scrope, speaking of the solidified streams of basalt, in the volcanic district of Southern France, observes :—

" These ancient currents have since been corroded by rivers, which have worn through a mass of 150 feet in height, and formed a channel even in the granite rocks beneath, since the lava first flowed into the valley. In another spot, a bed of basalt, 160 feet high, has been cut through by a mountain stream. The vast excavations effected

---

* " It is now admitted by all competent persons, that the formation even of those strata which are nearest the surface, must have occupied vast periods, probably millions of years, in arriving at their present state."—BABBAGE, Ninth Bridgewater Treatise, p. 67.

by the erosive power of currents along the valleys which feed the Ardèche, since their invasion by lava-currents, prove that even the most recent of these volcanic eruptions belong to an era incalculably remote." *

6. A series of organic beings appears, lives, generates, dies ; lives, generates, dies ; for thousands and thousands of successive generations. Tiny polypes gradually build up gigantic masses of coral, —mountains and reefs—microscopic foraminifera accumulate strata of calcareous sand; still more minute infusoria—forty millions to the inch—make slates, many yards thick, of their shells alone.

7. The species at length die out—a process which we have no data to measure,† though we may rea-

---

* Geology of Central France.

† " Though perfect knowledge is not possessed, yet there are reasons for believing that the duration of life to testacean individuals of the present race is several years. But who can state the *proportion* which the average length of life to the individual mollusc or conchifer, bears to the duration appointed by the Creator to the species ? Take any one of the six or seven thousand known recent species; let it be a *Buccinum*, of which 120 species are ascertained, (one of which is the commonly known *whelk ;*) or a *Cyprœa*, comprising about as many, (a well-known species is on almost every mantel-piece, the *tiger-cowry ;*) or an *Ostrea* (*oyster*), of which 130 species are described. We have reason to think that the individuals have a natural life of at least six or seven years; but we have no reason to suppose that any one species has died out, since the Adamic creation. May we then, for the sake of an illustrative argument, take the

sonably conclude it very long. Sometimes the whole existing fauna seems to have come to a sudden violent end; at others, the species die out one by one. In the former case suddenly, in the latter progressively, new creatures supply the place of the old. Not only do species change; the very genera change, and change again. Forms of beings, strange beings, beings of uncouth shape, of mighty ferocity and power, of gigantic dimensions, come in, run their specific race, propagate their kinds generation after generation,—and at length die out and disappear; to be replaced by other species, each approaching nearer and nearer to familiar forms.

8. Though these early creatures were unparalleled by anything existing now, yet they were animals of like structure and economy essentially. We can determine their analogies and affinities; appoint them their proper places in the orderly plan of nature, and show how beautifully they fill

duration of testacean species, one with another, at one thousand times the life of the individual? May we say six thousand years? We are dealing very liberally with our opponents. Yet in examining the vertical evidences of the cessations of the fossil species, marks are found of an entire change in the forms of animal life; we find such cessations and changes to have occurred MANY times in the thickness of but a few hundred feet of these late-rocks."—DR. J. PYE SMITH, *Scripture and Geology*, 5th Ed. p. 376.

hiatuses therein. They had shells, crusts, plates, bones, horns, teeth, exactly corresponding in structure and function to those of recent animals. In some cases we find the young with its milk-teeth by the side of its dam with well-worn grinders. The fossil excrement is seen not only dropped, but even in the alimentary canal. Bones bear the marks of gnawing teeth that dragged them and cracked them, and fed upon them. The foot-prints of birds and frogs, of crabs and worms, are imprinted in the soil, like the faithful impression of a seal.*

9. Millions of forest-trees sprang up, towered to heaven, and fell, to be crushed into the coal strata which make our winter fires. Hundreds

---

* "One of the laminated formations [in Auvergne] may be said to furnish a chronometer for itself. It consists of sixty feet of siliceous and calcareous deposits, each as thin as pasteboard, and bearing upon their separating surfaces the stems and seed-vessels of small water-plants in infinite numbers ; and countless multitudes of minute shells, resembling some species of our common snail-shells. These layers have been formed with evident regularity, and to each of them we may reasonably assign the term of one season, that is a year. Now thirty of such layers frequently do not exceed one inch in thickness. Let us average them at twenty-five. The thickness of the stratum is at least sixty feet; and thus we gain, for the whole of this formation alone, eighteen thousand years."—DR. J. P. SMITH, *Scripture and Geology*, 5th Edition, p. 137.

of feet measure the thickness of what were once succulent plants, but pressed together like paper-pulp, and consolidated under a weight absolutely immensurable. Yet there remain the scales of their stems, the elegant reticulated patterns of their bark, the delicate tracery of their leaf-nerves, indelibly depicted by an unpatented process of "nature-printing." And when we examine the record,—the forms of the leaves, the structure of the tissues, we get the same result as before, that the plants belonged to a flora which had no species in common with that which adorns the modern earth. Very gradually, and only after many successions, not of individual generations, but of the cycles of species, genera, and even families, did the vegetable creation conform itself to ours.*

10. At length the species both of plants and

---

* "This fact has now been verified in almost all parts of the globe, and has led to a conviction that at successive periods of the past the same area of land and water has been inhabited by species of animals and plants as distinct as those which now people the antipodes, or which now co-exist in the arctic, temperate, and tropical zones. It appears that from the remotest periods there has been ever a coming in of new organic forms, and an extinction of those which pre-existed on the earth; some species having endured for a longer, others for a shorter time; but none having ever re-appeared, after once dying out."—LYELL'S *Elements of Geology*, p. 275.

animals grew,—not by alteration of their specific characters, but by replacement of species by species—more and more like what we have now on the earth, and finally merged into our present flora and fauna, about the time when we find the first geological traces of MAN.

11. During the course of these successive cycles of organic life, the map of the world has changed many times. Up to a late period the ocean washed over Mont Blanc and Mount Ararat; the continent of Europe was a wide sea; then it was a Polynesia, then an Archipelago of great islands, then a Continent much larger than it is now, with England united to it, and the solid land stretching far away into the Atlantic;—then it sank again, and was again raised, not all at once, but by several stages, each of which has left its coast line, and its shingle beach. All these changes must have been the work of vast periods of time.

" Excepting possibly, but not certainly, the higher parts of some mountains, which at widely different epochs have been upheaved, and made to elevate and pierce the stratified masses which once lay over them, there is scarcely a spot on the earth's surface which has not been many times in succession the bottom of the sea, and a portion of dry land. In the majority of cases, it is shown, by physical evidences of the most decisive kind, that each of those successive conditions was of extremely long duration; a duration which it would be presumptuous to put into any estimate of years or centuries; for any alteration, of which vestiges occur in the zoological state and the mineral constitution of the earth's present surface, furnishes no analogy (with regard to the nature and continuance of causes), that approaches in greatness of character to those changes whose evidences are discernible in almost any two continuous strata. It is an inevitable inference, unless we are disposed to abandon the principles of fair reasoning, that each one of such changes in organic life did not take place till after the next preceding condition of the earth had continued through a duration, compared with which six thousand years appear an inconsiderable fraction of time." *

12. The climate of our atmosphere has undergone corresponding mutations. At one time the Palms, the Treeferns, the Cycads of the tropical jungles found their congenial home here : the Elephant, the Rhinoceros, and the Tiger roamed over England; nay, dwelt in countless hosts on the

* J. Pye Smith, Scripture and Geology, 5th Ed., p. 69.

F

northern shores of Siberia: then the climate gradually cooled to a temperate condition: then it
became cold, and glaciers and icebergs were its
characteristic features: finally it became temperate
again.

13. The icebergs and the glaciers were the ships
and railways of past epochs; they were freighted
with their heavy but worthless cargoes of rock-
boulders and gravel, and set out on their long
voyages and travels, over sea and land, sometimes
writing their log-books in ineffaceable scratches on
the rocky tables over which they passed, and at
length discharging their freights in harbours and
bays, on inland plains, on mountain sides and
summits, where they remain unclaimed, free for
any trader in such commodities, without the cere-
mony of producing the original bill of lading.

Let the remainder be told in the words of one of
our most eloquent and able geologists, Professor
Sedgwick.

" The fossils demonstrate the time to have been
long, though we cannot say *how* long. Thus we
have generation after generation of shell-fish, that
have lived and died on the spots where we find
them ; very often *demonstrating* the lapse of *many
years* for a few perpendicular inches of deposit.

In some beds we have large, cold-blooded reptiles, creatures of long life. In others, we have traces of ancient forests, and enormous fossil trees, with concentric rings of structure, marking the years of growth. Phenomena of this kind are repeated again and again; so that we have three or four distinct systems of deposit, each formed at a distinct period of time, and each characterised by its peculiar fossils. Coeval with the Tertiary masses, we have enormous lacustrine deposits; sometimes made up of very fine thin laminæ, marking slow tranquil deposits. Among these laminæ, we can find sometimes the leaf-sheddings and the insects of successive seasons. Among them also we find almost mountain-masses of the *Indusiæ tubulatæ* [the cases of *Phryganeæ*], and other sheddings of insects, year after year. Again, streams of ancient lava alternate with some of these lacustrine tertiary deposits.

" In central France, a great stream of lava caps the lacustrine limestone. At a *subsequent period* the waters have excavated deep valleys, cutting down into the lacustrine rock-marble many hundred feet; and, at a newer epoch, anterior to the authentic history of Europe, new craters have opened, and fresh streams of lava have run down the existing valleys. Even in the Tertiary period we have thus a series of demonstrative proofs of a long succession of physical events, each of which required a long lapse of ages for its elaboration.

" Again, as we pass downwards from the bottom Tertiary beds to the Chalk, we instantly find new types of organic life. The old species, which exist in millions of individuals in the upper beds, disappear, and new species are found in the chalk immediately below. This fact indicates a long lapse of time. Had the chalk and upper beds been formed simultaneously at the same period [as the supporters of the diluvial theory represent], their organic remains must have been more or less mixed; but *they are not.* Again, at the base of the Tertiary deposits resting on the Chalk, we often find great masses of conglomerate or shingle, made up of chalk-flints rolled by water. These separate the Chalk from the overlying beds, and many of the rolled flints contain certain petrified *chalk*-fossils. Now, every such fossil proves the following points :—

" 1. There was a time when the organic body was alive at the bottom of the sea.

" 2. It was afterwards imbedded in the cretaceous deposit.

" 3. It became petrified ; a very slow process.

" 4. The Chalk was, by some change of marine currents, washed away, or degraded, [*i. e.* worn away under the atmosphere by the weather and casualties, a process slow almost beyond description,] and the solid flints and fossils [thus detached from their imbeddings], were rolled into shingles.

" 5. Afterwards, these shingles were covered up, and buried under Tertiary deposits.

" In this way of interpretation, a section of *a few perpendicular feet* indicates a LONG lapse of time, and the co-ordinate fact of the entire change of organic types, between the beds above and those below, falls in with the preceding inference, and shows the lapse of time to have been VERY LONG."*

* In Dr. Pye Smith's Scripture and Geology, p. 382, (Ed. 1855.)

IV.

THE CROSS-EXAMINATION.

"When the fact itself cannot be proved, that which comes nearest to the proof of the fact is the proof of the circumstances that necessarily and usually attend such facts; and *these are called presumptions, and not proofs*, for they stand instead of the proofs of the fact, till the contrary be proved."—GILBERT; LAW OF EVIDENCE.

SUCH, then, is the evidence for the macro-chronology. I hope I have summed it up fairly; of course, many details I have been forbidden to adduce by want of space, but they would have been of the same kind as those brought forward. I am not conscious of having in any degree cushioned, or concealed, or understated a single proof which would have helped the conclusion.

A mighty array of evidence it certainly is, and such as appears at first sight to compel our assent to the sequent claimed for it. I must confess that I have no sympathy with the *reasonings* of those, however I honour their design, who can find a sufficient cause for these phenomena in the natural operations of the Antediluvian centuries, or in the convulsion that closed them.

But is there no other alternative? Am I compelled to accept the conclusions drawn from the phenomena thus witnessed unto, as undeniable facts, since they refuse to be normally circumscribed within the limits of the historic period? I verily believe there is another, and a perfectly legitimate solution.

My first business is to examine, and, if I can, to disprove this testimony. If I can show the witness to be liable to error; if I can adduce a principle which invalidates all his proofs ; if I can make it undeniably manifest that, in a case precisely parallel, similar conclusions, deduced from exactly analogous phenomena, would be notoriously false; if I can do this, I think I have a right to demand that the witness be bowed out of court, as perfectly nugatory and worthless *in this cause.*

In the first place, there is nothing here but *circumstantial* evidence; there is no *direct* testimony to the facts sought to be established. Let it not seem unfair to make this distinction; it is one of great importance. No witness has deposed to actual observation of the processes above enumerated; no one has appeared in court who declares he actually saw the living *Pterodactyle* flying about, or heard the winds sighing in the tops of the *Lepidodendra*. You will say, " It is the same thing; we have seen the skeleton of the one, and the crushed trunk of the other, and therefore we are as sure of their past existence as if we had been there at the time." No, it is not the same thing; it is not *quite* the same thing; NOT QUITE. Strong as is the evidence, it is not *quite* so strong as if you had actually seen the living things, and had been conscious of the passing of time while you saw them live. It is only by a process of reasoning that you infer they lived at all.*

* I would venture respectfully to suggest that the following argument by Mr. Babbage is vitiated throughout by a confounding of the phenomena observed with the conclusions inferred from them.

" What, then, have those accomplished, who have restricted the Mosaic account of the creation to that diminutive period, which is, as it were, but a span in the duration of the earth's existence, and who have imprudently rejected *the testimony of the senses*, when opposed to their philological criticisms? The very arguments which Protestants have opposed to the doctrine of transubstantiation, would, if their view of the case were correct, be equally irresistible against the Book of Genesis. But let us consider what would be the conclusion of any reasonable being in a parallel case. Let us imagine a manuscript written three thousand years ago, and professing to be a revelation from the Deity, in which it was stated that the colour of the paper of the very book now in the reader's hands is *black*, and that

The process is something like this. Here is an object in a mass of stone, which has a definite form,—the form of the bone of a beast. The more minutely you examine it, the more points of resemblance you find; you say, If this is a bone, it ought to have so and so—condyles, scars for the attachment of muscles in particular spots, a cavity for the reception of marrow, a mark for the insertion of the ligament; you look for each of these, and find all in the very conditions you have prescribed; it is not only a bone, but a particular bone, the thigh-bone, for instance. Here in the same block of stone is another object: you work it out; it is another bone; its joint accurately fits the preceding; it answers precisely to the tibia of a mammal. Other bones at length appear, and you have got a perfect skeleton, no part redundant, none wanting; the most minute, the most elaborate, the most delicate portions of the osseous frame

the colour of the ink in the characters which he is now reading is *white*. With that reasonable doubt of his own individual faculties which would become the inquirer into the truth of a statement said to be derived from so high an origin, he would ask all those around him, whether to their senses the paper appeared to be *black*, and the ink to be *white*. If he found the senses of other individuals agree with his own, then he would undoubtedly pronounce the alleged revelation a forgery, and those who propounded it to be either deceived or deceivers."— *Ninth Bridgewater Treatise*, p. 68.

of a mammal are present, and every one exactly correspondent to the rest in size, in maturity, *in fit.* Each bone, out of the scores, displays exactly those characters, and no other, which an anatomist would have said beforehand it ought to have. Allowing for the difference of species, the skeleton, when worked out of its matrix, and set up, is precisely like that of the little beast at whose death you were actually present, whose bones you cleaned with your own hands, and mounted for your own museum. It would be as reasonable to deny that the one is the skeleton of a real animal as the other.

Thus far there is matter of fact—observed, witnessed fact; you have found in a stone a real skeleton.

You immediately infer that this skeleton once belonged to a living animal, that breathed, and fed, and walked about, exactly as animals do now. This conclusion seems so obvious and unavoidable, that we naturally conclude it to rest on the same foundation as the fact that the object *is* a skeleton, or that *it was* in the stone. But really it rests on a totally different foundation; it is a conclusion deduced by a process of reasoning from certain assumed premises.

Myriads, perhaps millions of skeletons of animals like this one have come at different times under human observation, which have been obviously referrible to creatures that, within the same sphere of observation, had been alive. No similar skeleton has ever come within the range of recorded observation that could be referred to any other source than that of a quondam living animal. On these premises you build the conclusion that a skeleton must, at some time or other, have belonged to a living animal. And it may seem an impregnable position; but yet its validity altogether depends on the exhaustive power of human observation. If I could show, to your satisfaction, that a skeleton might have existed; still more, if I could show you that a skeleton *must* have existed; still more, if I could prove that myriads of skeletons, precisely like this, must have existed, without ever having formed parts of antecedent living bodies; you would yourself acknowledge that your conclusions were untenable. The utmost you could affirm, would be, that possibly, perhaps probably, the skeleton you had found in the stone had at some time belonged to a living animal, but that, so far as any recognised premises exist, there was no certainty about it.

But the premises have not been fairly stated. There is more than the relation of precedence and sequence in what we know of the connexion between skeletons and living animals ; there is the relation of cause and effect. It is not only that universal experience has declared the *fact* that every skeleton was once part of a living body; it has shown that the very structure and nature of the skeleton *implied* the living body. The skeleton, in every part, displays a regard for the advantages of the living animal ; it is built expressly for it ; by itself it is nothing—a machine without any object ; its joints, its cavities, its apophyses, its processes, all have special reference to tissues and organs which are not here now, but which belong to the living body.

And then experience has shown that the skeleton is made in a particular manner. The bone is deposited, atom by atom, in living organic cells, which are formed by living blood, which implies a living animal. The microscopic texture of your stone-girt skeleton does not differ from that of the skeleton which you cleaned from the muscles with your own hands; and therefore you infer that it was constructed in the same way, namely, by the blood of a living body.

Well, I come back, notwithstanding, to my position,—that your right to *affirm* this must altogether depend on the exhaustive power of that experience on which you build. And it will be overthrown, if I can show that skeletons have been made in some other way than by the agency of living blood.

Can I do this? I think I can. At least I think I can show enough greatly to diminish, if not altogether to destroy, the confidence with which you inferred the existence of vast periods of past time from geological phenomena. I can adduce a principle, having the universality (within its proper sphere) of LAW, hitherto unrecognised, whose tendency is to invalidate the testimony of your witness.

V.

POSTULATES.

" A little philosophy inclineth a man's mind to atheism; but depth in philosophy bringeth men's minds about to religion; for while the mind of man looketh upon second causes scattered, it may sometimes rest in them, and go no farther; but when it beholdeth the chain of them confederate and linked together, it must needs fly to Providence and Deity."—BACON.

" ' What was the opinion of Pythagoras concerning wildfowl?'
' That the soul of our grand-dam might haply inhabit a bird.'
' What thinkest thou of his opinion?'
' I think nobly of the soul, and in nowise receive his opinion.'"
                                                    SHAKSPEARE.

As without some common ground it is impossible to reason, I shall take for granted the two following principles :—

I. THE CREATION OF MATTER.
II. THE PERSISTENCE OF SPECIES.

I. If any geologist take the position of the necessary eternity of matter, dispensing with a Creator, on the old ground, *ex nihilo nihil fit,*— I do not argue with him. I assume that at some period or other in past eternity there existed nothing but the Eternal God, and that He called the universe into being out of nothing.

II. I demand also, in opposition to the development hypothesis, the perpetuity of specific characters, from the moment when the respective creatures were called into being, till they cease to be. I assume that each organism which the Creator educed was stamped with an indelible specific character, which made it what it was, and distinguished it from everything else, however near or like. I assume that such character has been, and is, indelible and immutable; that the characters which distinguish species from species *now*, were as definite at the first instant of their creation as now, and are as distinct now as they were then. If any choose to maintain, as many do, that species were gradually brought to their present maturity from humbler forms,—whether by the force of appetency in individuals, or by progressive development in generations—he is welcome to his hypothesis, but I have nothing to do with it. These pages will not touch him.

I believe, however, there is a large preponderance of the men of science,* at least in this country,

* Dr. Pye Smith calls the hypothesis of progressive development " the crude impertinence of a few foreign sophists,"—and

who will be at one with me .here.   They acknowledge the almighty *fiat* of God, as the energy which
produced being; and they maintain that the specific character which He then stamped on his
organic creation remains unchangeable.

he states as a fact, "that all the great geologists repudiate such
a notion with abhorrence, and give physical evidence of its false-
hood."—*Scripture and Geology*, (5th Ed.) p. 420.   See also
Professor Owen in "Rep. Brit. Assoc." 1842; Professor Sedgwick, in "Discourse on Stud. of Camb.;" Professor Whewell,
in "Hist. of Inductive Sciences;" Professor Ansted, in "Anc.
World;" &c.

VI.

LAWS.

" —— τὸν τροχὸν τῆς γενέσεως."—JAMES iii. 6.

THE course of nature is a circle. I do not mean the *plan* of nature; I am not speaking of a circular arrangement of species, genera, families, and classes, as maintained by MacLeay, Swainson, and others. Their theories may be true, or they may be false; I decide nothing concerning them; I am not alluding to any *plan* of nature, but to its *course*, *cursus*,—the way in which it *runs on*. This is a circle.

Here is in my garden a scarlet runner. It is a slender twining stem some three feet long, beset with leaves, with a growing bud at one end, and with the other inserted in the earth. What was it a month ago? A tiny shoot protruding from between two thick fleshy leaves scarcely raised above the ground. A month before that? The thick fleshy leaves were two oval cotyledons, closely appressed face to face, with the minute plumule between them, the whole enclosed in an unbroken, tightly-fitting, spotted, leathery coat. It was a bean, a seed.

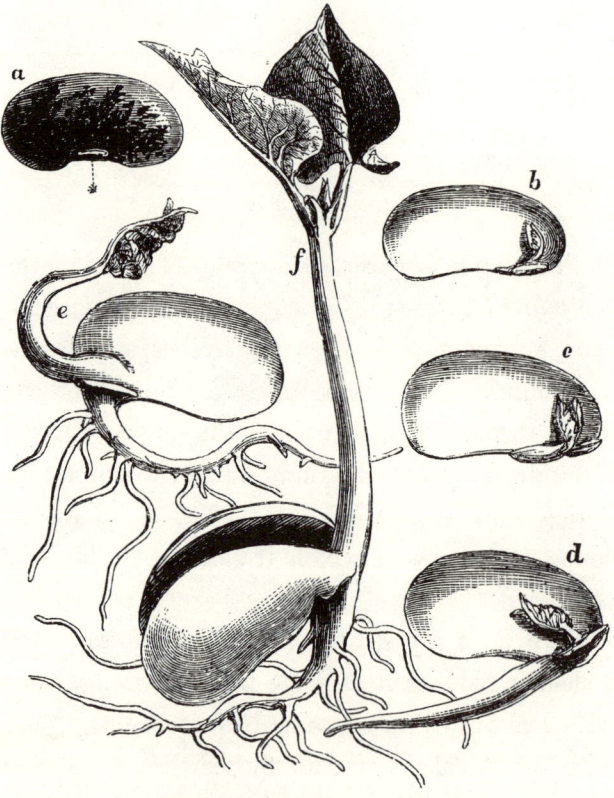

GERMINATION OF A SCARLET RUNNER.

a. The ripe bean, showing the hilum at * ;
b. The same bean, with one cotyledon removed, to show the plumule.
c. A similar bean, twenty-four hours after planting.
d. The same, on the sixth day after planting.
e. The same, on the twelfth day.
f. The same, on the fourteenth day.

N.B. From *b, c, d, e,* the front cotyledon has been cut away, to show the progress of the plumule.

Was this the commencement of its existence? O no! Six months earlier still it was snugly lying, with several others like itself, in a green fleshy pod, to the interior of which it was organically attached. A month before that, this same pod with its contents was the centre of a scarlet butterfly-like flower, the bottom of its pistil, within which, if you had split it open, you would have discerned the tiny beans, whose history we are tracing backwards, each imbedded in the soft green tissue, but no bigger than the eye of a cambric needle.

But where was this flower? It was one of many that glowed on my garden wall all through last summer; each cluster springing as a bud from a slender twining stem, which was the exact counterpart of that with which we commenced this little life-history.

And this earlier stem,—what of it? It too had been a shoot, a pair of cotyledons with a plumule, a seed, an integral part of a carpel, which was a part of an earlier flower, that expanded from an earlier bud, that grew out of an earlier stem, that had been a still earlier seed, that had been— and backward, *ad infinitum*, for aught that I can perceive.

The course, then, of a scarlet runner is a circle,

without beginning or end:—that is, I mean, without a natural, a normal beginning or end. For at what point of its history can you put your finger, and say, " Here is the commencement of this organism, before which there is a blank ; here it began to exist?" There is no such point ; no stage which does not look back to a previous stage, on which *this* stage is inevitably and absolutely dependent.

To some of my readers this may be rendered more clear by the accompanying diagram :—

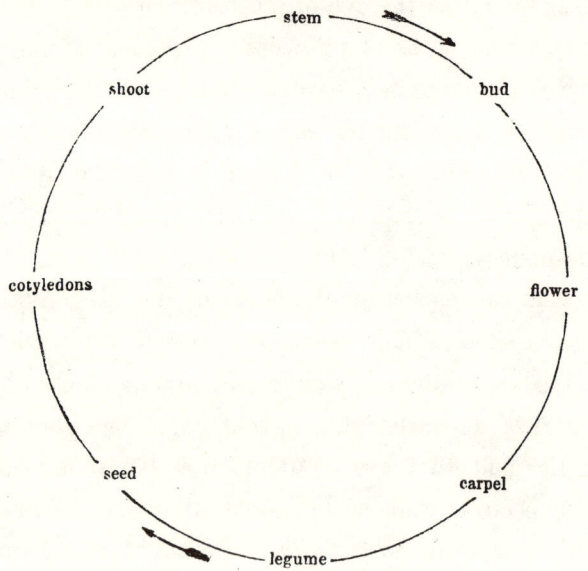

See that magnificent tuft of Lady-fern on yonder bank, arching its exquisitely cut fronds so elegantly on every side. A few years ago this ample crown was but a single small frond, which you would probably not have recognised as that of a Lady-fern. Somewhat earlier than this, the plant was a minute flat green expansion (*prothallus*), of no definite outline, very much like a Liverwort. This had been previously a three-sided spore lying on the damp earth, whither it had been jerked by the rupture of a capsule (*theca*). For this spore, though so small as to be visible only by microscopic aid, had a previous history, which may be traced without difficulty. It was generated with hundreds

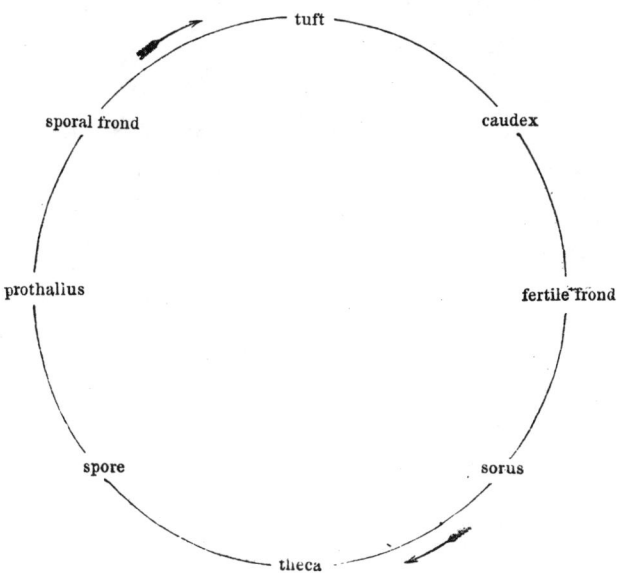

more, in one of many capsules, which were crowded together, beneath the oval bit of membrane, that covered one of the brown spots (*sori*), which were developed in the under surface of the fronds of an earlier Lady-fern. That earlier individual had in turn passed through the same stages of sporal frond, prothallus, spore, theca, sorus, frond, prothallus, spore, theca, sorus, frond, prothallus, &c.—*ad infinitum.*

This sounding-winged Hawkmoth, which like a gigantic bee is buzzing around the jasmine in the deepening twilight, hovering ever and anon to probe the starry flowers that make the evening air almost palpable with fragrance,—this moth, what " story of a life " can he tell? Nearly a year of existence he has spent as a helpless, almost motionless pupa, buried in the soft earth, from whence he has emerged but this evening. About a twelvemonth ago he was a great fat green caterpillar with an arching horn over his rump, working ever harder and harder at devouring poplar leaves, and growing ever fatter and fatter. But before that he had one day burst forth a little wriggling worm, from a globular egg glued to a leaf. Whence came the egg? It was developed within the ovary of a parent Hawkmoth, whose history is but an endless rotation of the same stages,—pupa, larva, egg, moth, pupa, larva, &c. &c.

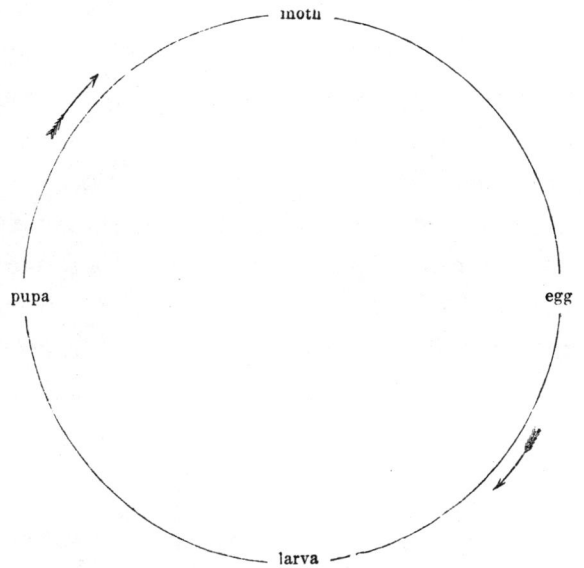

Behold this specimen of *Plumularia*, a shrub-like zoophyte, comprising within its populous branches some twenty thousand polypes. Every individual cell, now inhabited by its tentacled Hydra, has in its turn budded out from a branch, which was itself but a lateral process from the central axis. And this was but the prolongation of what was at first a single cell, shooting up from a creeping root-thread. A little earlier than this, there was neither cell nor root-thread, but the organism existed in the form of a *planule*, a minute soft-bodied, pear-shaped worm, covered with cilia, that crawled slowly over the stones and sea-weeds. Whence came it? A few hours before, it had emerged from the mouth of a vase-like cell, one of the ovarian capsules, which studded the stem of an old well-peopled Plumularia-shrub, and which had been gradually developed from its substance by a process analogous to budding. And then if we follow the history of this earlier shrub backward, it will only lead us through exactly correspondent stages, primal cell, planule, ovarian capsule, stem, and so on interminably.

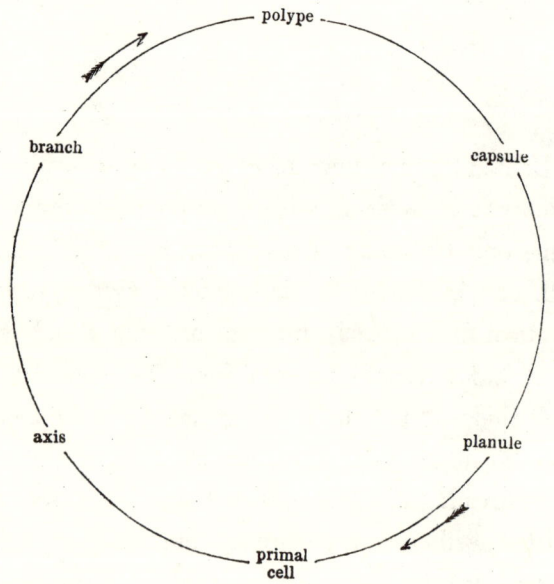

Once more. The cow that peacefully ruminates under the grateful shadow of yonder spreading beech, was, a year or two ago, a gamesome heifer with budding horns. The year before, she was a bleating calf, which again had been a breathless fœtus wrapped up in the womb of its mother. Earlier still it had been an unformed embryo; and yet earlier, an embryonic vesicle, a microscopically minute cell, formed out of one of the component cells of a still earlier structure,—the germinal vesicle of a fecundated ovum. But this ovum, which is the remotest point to which we can trace

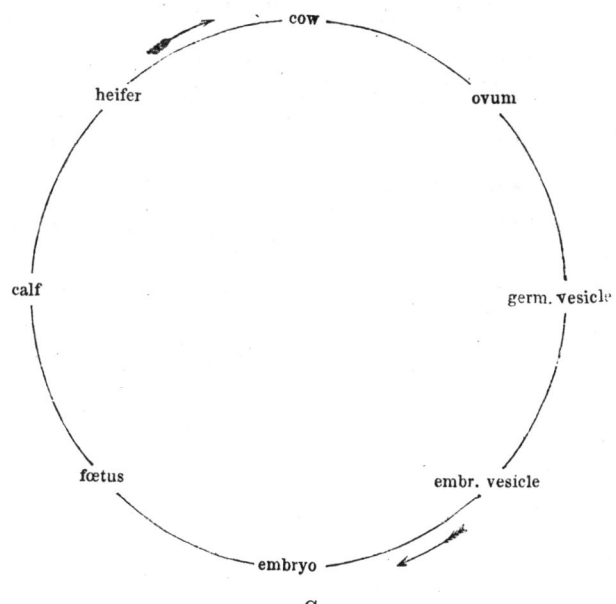

the history of our cow as an individual, was, before it assumed a distinct individuality, an undistinguishable constituent of a viscus,—the ovary,—of another cow, an essential part of *her* structure, a portion of the tissues of *her* body, to be traced back, therefore, through all the stages which I have enumerated above, to the tissues of another parent cow, thence to those of a former, and so on, through a vista of receding cows, as long as you choose to follow it.

This, then, is the order of all organic nature. When once we are in any portion of the course, we find ourselves running in a circular groove, as endless as the course of a blind horse in a mill. It is evident that there is no one point in the history of any single creature, which is a legitimate beginning of existence. And this is not the law of some particular species, but of all : it pervades all classes of animals, all classes of plants, from the queenly palm down to the protococcus, from the monad up to man : the life of every organic being is whirling in a ceaseless circle, to which one knows not how to assign any commencement,—I will not say any certain or even probable, but any *possible*, commencement. The cow is as inevitable a sequence of the embryo, as the embryo is of the cow.

Looking only at nature, or looking at it only with the lights of experience and reason, I see not how it is possible to avoid one of these two theories, the development of all organic existence out of gaseous elements, or the eternity of matter in its present forms.

Creation, however, solves the dilemma. I have, in my postulates, begged the fact of creation, and I shall not, therefore, attempt to prove it. Creation, the sovereign fiat of Almighty Power, gives us the commencing point, which we in vain seek in nature. But what is creation? It is *the sudden bursting into a circle.* Since there is no one stage in the course of existence, which more than any other affords a natural commencing point, whatever stage is selected by the arbitrary will of God, must be an un-natural, or rather a preter-natural, commencing point.

The life-history of every organism commenced at some point or other of its circular course. It was created, called into being, in some definite stage. Possibly, various creatures differed in this respect; perhaps some began existence in one stage of development, some in another; but every separate organism had a distinct point at which it began to live. Before that point there was nothing; this particular organism had till then no existence; its history presents an absolute blank; *it was not.*

But the whole organisation of the creature thus newly called into existence, looks back to the course of an endless circle in the past. Its whole structure displays a series of developments, which as distinctly witness to former conditions as do those which are presented in the cow, the butterfly, and the fern, of the present day. But what former conditions? The conditions thus witnessed unto, as being necessarily implied in the present organisation, were non-existent; the history was a perfect blank till the moment of creation. The past conditions or stages of existence in question, can indeed be as triumphantly inferred by legitimate deduction from the present, as can those of our cow or butterfly; they rest on the very same evidences; they are identically the same in every respect, except in this one, that they were *unreal.* They exist only in their results; they are effects which never had causes.

Perhaps it may help to clear my argument if I divide the past developments of organic life, which are necessarily, or at least legitimately, inferrible from present phenomena, into two categories, separated by the violent act of creation. Those unreal developments whose apparent results are seen in the organism at the moment of its creation, I will call *prochronic*, because time was not an element in them; while those which have subsisted since creation, and which have had actual existence, I will distinguish as *diachronic*, as occurring during time.

Now, again I repeat, there is no imaginable difference to sense between the prochronic and the diachronic development. Every argument by which the physiologist can prove to demonstration that yonder cow was once a fœtus in the uterus of its dam, will apply with exactly the same power to show that the newly created cow was an embryo, some years before its creation.

Look again at the diagram by which I have represented the life-history of this animal. The only mode in which it can begin is by a sudden sovereign act of power, an irruption into the circle. You may choose *where* the irruption shall occur; there must be a bursting-in at some point. Suppose it is at " calf;" or suppose it is at " embr. vesicle." Put a wafer at the point you choose, say the latter. This then is the real, actual commencement of a circle, to be henceforth ceaseless. But the embry-onic vesicle necessarily implies a germinal vesicle, and this necessitates an ovum, and the ovum necessitates an ovary, and the ovary necessitates an entire animal,—and thus we have got a quarter round the circle in back development; we are irresistibly carried along the prochronic stages,— the stages of existence which were before existence commenced,—as if they had been diachronic, actually occurring within our personal experience.

If I know, as a *historic fact*, that the circle was commenced where I have put my wafer, I may begin it there; but there is, and can be, nothing in the *phenomena* to indicate a commencement there, any more than anywhere else, or, indeed, anywhere at all. The commencement, as a fact, I must learn from testimony; I have no means whatever of inferring it from phenomena.

Permit me, therefore, to repeat, as having been proved, these two propositions :—

ALL ORGANIC NATURE MOVES IN A CIRCLE.

CREATION IS A VIOLENT IRRUPTION INTO THE CIRCLE OF NATURE.

VII.

PARALLELS AND PRECEDENTS.

(*Plants.*)

"Where wast thou when I laid the foundations of the earth? declare, if thou hast understanding."—JOB xxxviii. 4.

SINCE every organism, considering it, throughout its generations, as an unit, has been created, or made to commence existence, it is manifest that it was created or made to commence existence at some moment of time. I will ask some kind geological reader to imagine that moment, and to accompany me in an ideal tour of inspection among the creatures, taking up each for examination at the instant that it has been called into existence. Do not be alarmed! I am not about to assume that the moment in question was six thousand years ago, and no more; I will not rule the actual date at all; you, my geological friend, shall settle the chronology just as you please, or, if you like it better, we will leave the chronological date out of the inquiry, as an element not relevant to it. It may have been six hundred years ago, or six thousand, or sixty times six millions; let it for the present remain an indeterminate quantity. Only please to remember that the date *was* a reality, whether we can fix it or not; it *was* as precise a moment as the moment in which I write this word.

Well then, like two of those " morning stars " who, when " the foundations were fastened," " shouted for joy," we will, in imagination, take our stand on this round world at exactly —— minutes past —— o'clock, on the morning of the ——th of ——, in the year B.C. ——. The noble Tree-fern before us (*Alsophila aculeata*) has this instant been called into being by the creating voice of God. Here it stands, lifting up its columnar stem, and spreading its minutely fretted fronds all around, in a vaulted canopy above our heads, through the filagree work of whose expanse the sunbeams play in a soft green radiance. It has this instant been created.

But I will suppose, further, that we have the power to call into our council some experienced botanist; who is not acquainted, as we are, with the fact of this just recent creation, and whom we will

ask to give us his opinion on the age of this
beautiful plant.

The Botanist.—" You wish to ascertain the age
of this *Alsophila*. I know of no data by which
this can be determined with precision, but I can
indicate it approximately. Let us take it in order.
The most recent development is the growing point
in the centre of the arching crown of leaves.
Around this you would see, if your eyes were
above the plane, close ring-like bodies, or, perhaps,
more like snail-shells, protruding from the growing
bud; then young leaves, partially opened in various
degrees, but coiled up scroll-wise at their tips, and
around these the elegant fretted fronds, which
expand broadly outwards in a radiating manner,
and arch downwards.

" Now every one of these broad fronds was at
first a compactly coiled ring; but it has, in the
course of development, uncoiled itself, growing at
the same time from its extremity, and' from the
extremity of each of its formerly wrapped-up
pinnæ and pinnules, until at length it has attained
the expanse you behold. This process has certainly occupied several days.

" But let us look farther. The outermost fronds
that compose this exquisite cupola, you see, are

nearly naked; indeed, the extreme outermost are
quite naked, being stripped of their verdant honours,
their pinnæ and pinnules, and left mere dry and
sapless sticks,—the long and taper midribs of
what were once green fronds, as graceful as those
that now surmount them. Some of them, you
see, are hanging downward, almost detached
from the stem, and ready to drop at the first
breath of wind. Now remember, each of these
brown unsightly sticks was once a frond, that had
passed through all the steps of uncoiling from its
circinate condition. This whole process has certainly occupied several months.

" Look, now, below these withered midribs,
lifting up the most drooping of them. The stem
is marked with great oval scars; and see, this old
frond-rib has come off in my hand, leaving just
such a scar, and adding one more to the number
that were there before. And look down the stem;
it is studded all over with these oval scars. There
are a hundred and fifty at least; but I cannot
count them nearly all, for towards the lower part
they become more undefined, and the growth of
the stem has thrown them further apart; and
besides, there is, as you observe, a matted mass of
tangled rootlets, like tarred twine, which, springing

from between the lower scars, increases downwards, till the whole inferior extremity of the stem is encased in the dank and reeking mass.

"You can have no doubt that every one of these scars indicates where a leaf has grown, where it has waved its time, and whence, after death and decay, it at length sloughed away. The form of the uppermost, which are not distorted by age, agrees exactly with the outline of the bulging base of the candelabrum-like frond; the arrangement of the scars is that of the fronds; and you may notice in every scar marks where the horseshoe-shaped plates of woody fibre have been broken off, which once passed into the interior of the stem from the midrib of the frond.

"These scars, then, are ocular demonstrations of former fronds; we may no more doubt that fronds were once growing from these spots, than we may that the green and leafy arches were once coiled up in a circinate vernation. They are the record of the past history of this organism, and they evidently reach far back into time. The periodic ratio of development of new fronds may be, perhaps, roughly estimated at six in the course of a year. Now there are about a dozen unfolded or unfolding, as many withering midribs, and about a hundred and fifty leaf-scars that we can count with ease, not reckoning such as are indistinct, nor such as are concealed beneath the tangled drapery of roots.

LEAF-SCARS OF TREE-FERN.

"I have no hesitation, then, in pronouncing this plant to be thirty years old; it is probably much older, but it is, at least, as old as this."

Such is the report of our botanical adviser; such is his argument; and we cannot but admit that it is invulnerable; his conclusion is inevitable, but for one fact, which he is not aware of. There *is* one objection, however, to which it is open—a fatal one; you and I know that the Tree-fern is not five minutes old, *for it was created but this moment.*

Here is another act of creation. It may be the same day as that of the Tree-fern, or one as remote as you please from it, before or after. A few moments ago this was a great mass of rough, naked limestone, but by creative energy it has been suddenly clothed with a luxuriant mantle of *Selaginella.* How exquisitely beautiful the aggregation of flattened branching stems, studded with their tiny imbricated leaflets of tender green, bloomed with blue! and how thick and soft the carpet that thus conceals the angles and points and crevices of the unsightly stone! Broad as is this expanse of verdure, covering many square yards without a flaw, and rooted as it is at ten thousand points of its creeping stem, we shall yet find that it is one unbroken structure. Our friend the botanist would infer unhesitatingly that every part of this widespread ramification has originally proceeded from one central shoot, and that several years' growth must have concurred to form this compact mass.

Yet *we* know that such an inference would be

false. The plant has been this instant called into being.

On the summit of this rounded hill is a very different plant from the last. Beautiful it also is, but grandeur and majesty are its leading attributes. It is a dense and massive clump of the Tulda Bamboo. How noble these straight-jointed stems, cylinders of polished green, shooting their points right upwards, and towering to a height of eighty feet! The numerous panicles of tufty blossom are gracefully bending from the summits, and from the tip of every branch, nodding in the breeze. There are scores of the tall stems, as straight as an arrow, beset at every joint with diverging horizontal branches, crossing and recrossing in inextricable confusion. And see, amidst the crowd, there are others as thick and tall, but without a single side-shoot, clothed, however, to atone for the deficiency, in swaddling-clothes peculiarly their own.

These swathed stems are infant shoots,—vigorous and promising children, indeed; these brown triangular sheaths, covered with down, are the clothing of infancy; they increase in number, and are closer together towards the summit of the shoot, where the growing point is rapidly extending. When the stems have attained their full height, these sheaths will fall off, the polished shafts will stand revealed in their glossy beauty, and the lateral pointed branches will at once start forth from every joint, and pierce horizontally through the dense tangled bush.

Now these young shoots do not bear testimony to so great an age as you would suppose. The whole seventy feet of their altitude have been attained within thirty days! But then their massive size and vigour indicate a mature age in the clump. For all the hundred stems that are crowded together in this dense Bamboo-clump are organically united; they are parts of one and the same plant, the root-stock of which has been creeping to and fro year after year, sending up in constant succession its arrowy stems, until it has attained the present magnificence. Many years must have elapsed between the present condition of the grove, and that of the slender blade that shot up from the tiny seed in this spot.

Yes, so you may think. But it is not so, for the great Bamboo-clump has been created in its pride and glory this very hour!

Yonder is a considerable area of land covered with the green blades of young wheat, and very healthy and strong it looks. No, it is Couch-grass!

The whole green sward which we see is a single plant; the creeping stem of which has spread its ramifications in all directions beneath the surface of the soil; and still the long succulent shoots are extending in every direction, as shewn by the green leaf-blades. This is a rapidly growing plant, it is true; yet still there must be an accumulated growth of many months here, if not years! No, it was created this morning.

Contrasting with this humble grass, observe that luxuriant Screwpine. See its singular crown of foliage at the summit of its equally singular stem. Its great prickle-edged stiff leaves grow in long diagonal rows, each sheathing its successor, and alternating with those of the next row. How rich and fragrant an odour is diffused from its crowded blossoms!

Every one of those sword-like leaves is, of course, the record of a period of time. A tree of this size makes a "screw," or imperfect spire, of leaves in about three years; and there are about sixteen pairs of leaves in each screw, which will give us nearly eleven leaves for the development of each season. Now, on the trunk, there are numerous waved lines quite covering its surface, which are the traces of old leaves that have in succession

been produced and decayed away;—the trunk is,
in fact, composed of these leaf-bases. By counting
these, we may obtain then an approximate notion
of the age of this plant;—an *approximate* notion
only, because in its young stages the development
of leaves probably took place more rapidly than it
does now. There are then on this trunk about one
hundred and fifty horizontal rows of scars, and each
row numbers four leaf-bases, so that the trunk is
inscribed with an autographic record of six hundred
leaves. If then we reckon eleven leaves as the
produce of a single season, and add the four screws
which are still flourishing, we shall obtain a result
of about fifty-five years as the age of this *Pandanus.*
This, for the reason just assigned, would probably
be considerably too much; perhaps, forty years
would be nearer the truth.

There are, however, other marks of age here,
though they are less definite. The great hardness
of the surface-wood, which we perceive on trying
to indent it, is an indication of age, as it is produced by the successive bundles of woody fibre,
which, year after year, have passed down from each
leaf, curving, in their descent, towards the circumference of the stem, and, therefore, constantly
augmenting the density of the outer portions.

Another very curious proof of age is seen in the number of aerial roots which descend from various points of the trunk towards the soil. You would at first be inclined to think them posts, which a carpenter had set to "shore up" the tree, as props to prevent its being blown down. And truly this is their purpose; but they are natural adjuncts, not artificial. These thick rods, some of which have not yet reached the ground, have been shot forth in turn from the stem, in order to afford it additional support in the loose sandy soil. And mark, by the way, a beautiful contrivance here. Because the growing tender extremity of the root has to pass through the sun-parched air in its progress towards the earth, there is a curious exfoliation of its extremity, forming a sort of cup, which, collecting the scanty dews, retains sufficient moisture for the refreshment of the spongy rootlet. Now, I say, these supporting roots, since they must have originated from the trunk, after the latter had attained a considerable height, are so many evidences—and cumulative evidences—of age, though their testimony cannot be so well made to bear on a known period as that of the leaf-bases.

Should we not then be amply warranted in asserting this Screw-pine to be many years old, if

we were not assured that, as a fact, it has been
this instant created?

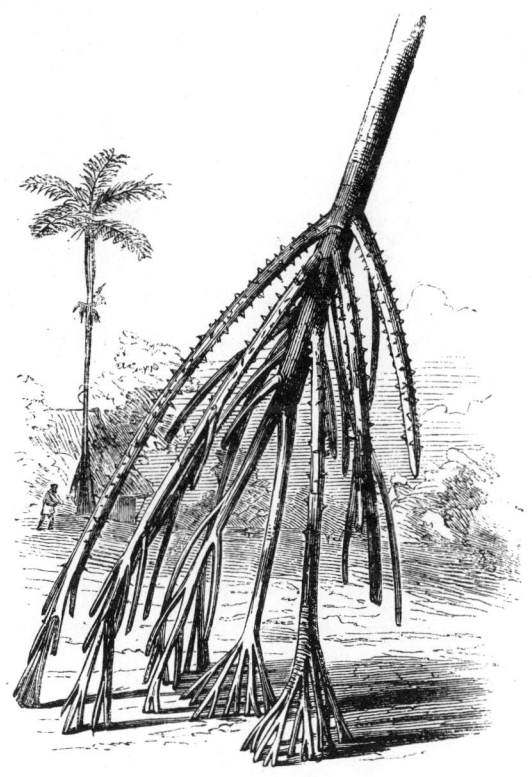

ROOTS OF IRIARTEA.

A phenomenon analogous to that which we have
just observed is presented by yonder Pashiuba
Palm *(Iriartea exorhiza)*. Its straight arrowy stem
has shot up to the height of fifty feet, like a slender

iron column. On the summit there is the usual divergent crown of leaves that distinguishes this most graceful and queenly tribe; and at the foot, a tall open cone of roots, strangely supporting the column on its apex.

"But what most strikes attention in this tree, and renders it so peculiar, is, that the roots are almost entirely above ground. They spring out from the stem, *each one at a higher point than the last*, and extend diagonally downwards till they approach the ground, when they often divide into many rootlets, each of which secures itself in the soil. As fresh ones spring out from the stem, *those below become rotten and die off;* and it is not an uncommon thing to see a lofty tree supported entirely by three or four roots, so that a person may walk erect beneath them, or stand with a tree seventy feet high growing immediately over his head."

"In the forests where these trees grow, numbers of young plants of every age may be seen, all miniature copies of their parents, except that they seldom possess more than three legs, which gives them a strange and almost ludicrous appearance."*

This tall Pashiuba before us, however, is sup-

* Wallace's "Palms of the Amazon," p. 35.

ported on several scores of roots, in various stages of development, some descending through the air, some already fixed in the soil. As the presence of these, moreover, implies the decay and disappearance of earlier ones, their number and height may be accepted as a fair testimony to the age of the tree; independent of what we might have deduced from the trunk and other sources. (My reader will bear in mind, that, throughout this chapter, I am supposing that we have the opportunity of seeing each organism at the moment following that of its creation.) The *Iriartea* before us, then, notwithstanding its marks of maturity, is but—a new-born infant, I was about to say, rather—a new-made adult.

Another and more massive Palm appears, where a moment ago there was nothing but smooth ground and empty air. It is the Sugar Palm *(Saguerus saccharifer)*, remarkable in its appearance for the swathes of what looks to be *sackcloth of hair*, in which its stem is enveloped. Each of its great pinnate leaves forms with the dilated base of its midrib a broad sheath, which springs out of a loose fold of this coarse cloth that is wrapped around it. And not only the bases of the still flourishing leaves are swathed in this natural textile fabric, but the dead and dry leaf-bases of the former leaves, which may be traced all down the stem. But let us look at this strange cloth : what is it? It is composed of the exterior fibres of the leaf-bases themselves, which in process of growth have partially separated themselves, and from which the parenchyma and the lamina have decayed away. The appearance of a woven fabric is deceptive; there is no interlacing; but its semblance is produced by the fibres lying in layers one over the other, and by some of them having a direction at right angles to the others. Originally all the fibres were parallel and longitudinal, but as they have been, in the growth of the leaf, pulled out laterally, the main fibres, which are indefinitely divisible, have adhered to each other at various parts, and the result has been that innumerable constituent fibrillæ have been stretched across from fibre to fibre.

Every square inch, then, of this sackcloth tells of the lapse of time; these horse-hair-like fibres were once green and vascular, enclosing a soft pulp; in short, they were a part of a verdant leaf; the reduction of each congeries of veins to this condition was a work of time, and this has been effected by many leaf-bases in succession.

An examination of this *gomuti*, as it is called, does not indeed help us to identify the actual interval lapsed in the history of the plant; but we may arrive at this from other considerations. The great sheathing bases themselves remain in numbers attached to the upper portion of the stem, though the greater portion of the midrib with the pinnæ has decayed and fallen; and in the lower part, where even the bases have disappeared, still broad lateral scars are left, marking off the stipe into horizontal rings, which are not less conclusively certain evidences of the former existence of similar bases, and therefore, still earlier, of leaves.

The Sugar Palm developes and matures on an average six leaves every year.* On counting the dry leaf-bases, and the scars, I find on this trunk, a hundred and twenty: besides which there are about a dozen expanded leaves, and two visible, which are not unfolded. A hundred and thirty-four leaves then have left proofs of their existence here; which divided by six, gives about twenty-two years as the age of this Palm. This is the age of this tree, however, since it began to form a stem; but several years of infancy must be added to the

* Roxburgh.

sum, during which its fronds sprang in succession from the surface of the soil.

Look at this *Areca*. By-and-by it will grow to the loftiest stature attained by any of its tribe, and its noble crown of leaves will wave on the summit of a slender pillar a hundred and fifty feet in height. But at present it has no stem at all; the widely arching leaves diverge from a central point which is below the surface of the soil. Here, then, are no dead leaf-bases; here are no old historical scars:—have we any evidence of past time here? Yes, surely. See this fully developed leaf. It is composed of a stout midrib, along the two opposite edges of which grow, like the beards of a feather, narrow sword-like leaflets, separated from each other by intervals of about two inches. But this pinnate condition,—which is so inseparable from the developed leaf of a great division of the Palm tribe, that our idea of a palm-leaf almost always is that of an enormous feather,—is by no means the original state. Observe this young leaf which is not yet thoroughly expanded; the leaflets are, indeed, separated everywhere, except that the tips of all are connected by a very narrow ribbon of the common green lamina, which runs from one to another. In the fully opened leaves, this has been torn apart and is not distinguishable.

But, let us carefully open this still younger leaf, which is protruding like a thin green rod, or rather like a closed fan, from the centre of the crown. We must handle it delicately, for it is very tender. Now you see it is not pinnate at all; the leaf is as entire as a *Musa* leaf, which, indeed, it much resembles, except that each half is folded transversely, and then these transverse folds are packed one on another longitudinally, fan-fashion. Each of the transverse folds answers to a future leaflet. It is the development of the midrib in length that tears asunder the divisions of the lamina, and converts them into separate, and by-and-by remote, pinnæ.

It is manifest then that every leaflet on the midrib of a pinnate-leaved Palm is a record of past time, as real as the leaf-bases on the trunk, inasmuch as, in each case, there is ocular proof that the conditions of existence are different from what they have been. And yet in this case, the evidences are fallacious, since the *Areca* before us has even now been created.

Here is an extraordinary plant. Though no thicker than your little finger, it will be found almost a quarter of a mile in length.* This is a

* Rumph, v. 100.

H

kind of Cane (*Calamus*) ; its slender jointed and polished stem is encased in the closely-sheathing and tubular bases of the leaves, which are spiny on their midribs, spiny on their pinnæ, and horridly spiny on the long and tough whip-lash in which the point of each leaf terminates. This lengthened cord is studded, at intervals of a few inches, with whorls of stout and acute prickles which are hooked backwards, and performs an important part in the economy of the plant. We see how it sprawls along the ground a few yards, then climbs up a tall tree, runs over the summit, descends on the opposite side to the ground, mounts over another tree, and thus pursues its wormlike course. Now as the pinnate leaves are put forth at every joint, the formidably armed flagellum affords a secure holdfast to the climbing stem, which otherwise would be liable to be blown prostrate by the first gust of wind ; the recurved hooks, however, catch in the leaves and twigs of the trees, and effectually maintain the domination of the prickly intruder.

It is obvious that every inch sprawled over by this trailing stem supposes all the previous inches of its lengthening course; that every successive joint implies the existence of all the earlier joints ;

that every whorl of spines involves the development of every former whorl.   Yet our reasoning is
at fault; there has been as yet no succession; the
development has been simultaneous, for it is the
development, not of growth, but of creation.

Enough of Palms.   Look at this *Agave*.   Its
thick, fleshy, glaucous leaves, with spinous margins
and pointed ends, are arranged in many whorls on
the summit of a stem, which is scarcely visible, as
it barely rises above the soil.   From the centre of
the crown springs the stately flower-stalk, itself a
tree of forty feet in stature, having a cluster of
yellow blossoms at the extremities of its candelabra-
like branches.

Have we here any clue to the past history of the
plant?   The tall flower-stalk, it is true, is of rapid
growth, its whole stature having been attained
within three or four weeks.   But those massive
leaves!   Each of these lasts many years, and their
development is as slow as that of the flower-stalk
is rapid.   Certainly we cannot assign to this individual, in the very vigour of its inflorescence, an
antiquity less than half a century, and perhaps it
may be considerably more.

You are altogether wrong; for it is but just
called into existence.

We pass on, and pause before a noble example
of one of the stateliest of plants,—the Traveller's

TRAVELLER'S TREE.

Tree (*Urania speciosa*). It is a great Musaceous
plant, resembling one of those fans which in the

Southern States of America are made by ladies out of the broad tail-feathers of a turkey. Its leaves, of vast size, consist of a broad oblong lamina of the most brilliant green hue, divided equally by a midrib which descends in a smooth cylindrical petiole, much longer than the lamina (which is itself eight feet or more in length). Each leaf-stalk terminates below in a great demi-sheath, out of which springs another, in a zigzag or distichous fashion, the whole diverging, as they rise, in the same plane.

Below the alternately-sheathing leaves, of which there are but eight at present existing, there are the bases of others, now dead, which, when alive, evidently followed the same arrangement; and these give place yet lower to rings, each partly surrounding a massive conical stem.

I fear we have no criterion for determining the exact age of such a plant as this from actual observations on its rate of growth. From the fewness of its existing leaves they probably endure a considerable time; but at all events here are indubitable evidences of successive generations of leaves which are now past and gone; some of which are represented by withered rib-bases, while older ones have left but the scars which indicate the positions on the trunk where once they stood. Here are distinct testimonies to the lapse of a considerable period of time since the magnificent *Urania* began its existence. Yet we should err egregiously by giving credence to them, since these developments are all *prochronic.*

" What a lovely butterfly ! " Nay, it is a flower: though it dances in the air with an insect's fluttering flight, and seems to present in its broad wings of yellow and orange, and in its long and slender members, an insect's form and hues, it is but a flower fixed at the end of a lengthened stalk, which hangs from a mass of leaves and bulbs, seated in the fork of this huge mahogany-tree.

We will neglect the flower, curious and beautiful as it is, and examine this crowded mass of roots and fleshy leaves and oval bulbs.

Tracing the slender lengthened footstalk to its origin, we see that it springs from the lower part of a flat, ovate, or nearly round, ridged, pseudo-bulb, of a purplish-green hue, of which there are many, much crowded together. The point of issue of the flower-stalk is concealed by an enveloping husky scale, which is the withered condition of a former leaf. From the base of another bulb a thick obtuse cone is pushing forth, which is the com-

mencement of a new leaf-shoot; and here is one
considerably advanced. In this latter there is
nothing very remarkable; it is a thick growing
shoot, formed by fleshy leaves nearly doubled
together, each sheathed by its predecessor. But
soon this will cease to grow, and the point will
dilate into an oval bulb, which will be a reservoir
of nutriment for the future flower. In fact it will
add another to the matted mass of bulbs which are
already accumulated, crowned with two great thick,
leathery, ovate, brown-spotted leaves, and marked
with the scars of the leaves which are now growing,
but which will then have sloughed away.

In this *Oncidium*, then, we have evidently a record
of many bygone processes. Before the flower could
open, the flower-stalk must have been developed;
before this, the pseudo-bulb must have been formed;
before this, there must have been a well-formed
leaf-shoot, which must have been first a conical
bud pushing forth from some anterior bulb;—or,
if that shoot had been the first of the mass, then it
must have looked back to a seed, which of course
looked back to the capsule of a pre-existent flower,
and so on.

Yet this is all fallacious; for the Butterfly-flower
is but just created.

As beautiful, if less curious, is the crowded spike
of purple blossom that adorns the tall stalk of this
terrestrial Orchis. The flower-stalk springs from
the midst of a few large spotted leaves, which termi-
nate below in an irregular fleshy tuber of glutinous
consistence. This tuber is shrivelled, and is in
process of exhaustion and decay ; but a horizontal
stem has pushed out underground, which has at
its extremity a second tuber, as yet immature, but
plump and swelling. This growing tuber contains
the elements of the leaves and flower-spike of next
season : the shrivelling one was, last year at this
period, in exactly the same condition as the swelling one is now ; it too was pushed out horizontally
from a preceding one which was then shrivelling,
and so backward. These pre-existing stages can
with certainty be announced by the vegetable
physiologist ; who yet would be deceived in this
instance, because the plant has been but just
created.

This elegant *Gladiolus* that displays its tall spike
of crimson blossoms from the midst of its flattened
folded leaves, affords us a similar example of retrospective energy. If I dig away the light soil from
around its base, I discover two globose corms, fleshy
swellings of the stem, accumulations of nutriment

obtained during the vegetative activity of the plant, and destined to support it during the season of inaction, and therefore stored up for that purpose.

The uppermost of these globose corms is that of the present season; it is as yet small and immature, being in process of formation by the assimilation, consolidation, and deposition of new matter by the action of the leaves. This is sheathed

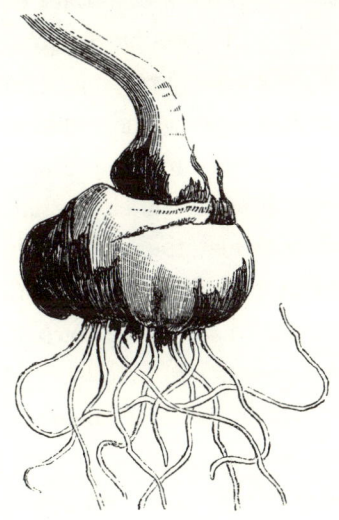

CORM OF GLADIOLUS IN JUNE.

in the tubular bases of the leaves, which. expand above; and it is seated on a larger, riper, and more spherical corm, which is wrapped in a brown fibrous skin. This is the matter which was deposited in the course of last spring and summer, and the brown skin is the remains of the leaves of last year. This corm has remained inactive, since the decay of last year's leaves, until this winter, when the root fibres, which we see descending from the lower surface, began to form, and an upward prolongation of the stem followed, which, as it grew, swelled into the upper corm.

In the centre of the under surface of the corm of last season, in a depression surrounded by the white root-fibres, there are some almost decayed remains of a deep brown hue. These are the last vestiges of the preceding year's corm, and they exhibit the condition in which the large corm will be next spring, when the small half-formed one will be in the state and position of this larger one, and will in like manner be surmounted by its rising successor.

Thus there are in this plant ocular proofs of two years' history before the present; yet these proofs are invalidated by the fact of its creation this day.

Behold now that singular plant, the Grass-tree (*Kingia australis*), displaying what seems an immense tuft of wiry grass elevated on the summit of a trunk which is formed of the united bases of myriads of decayed leaves, the representatives of

many generations of these organs. The silvery
leaves which constitute the existing crown, and
the numerous spikes of blossom which stand up in
a circle diverging from the midst of them, give to
this plant a most striking effect. That, however, is
not our present concern, but the evidences which
we may be able to gather from it of a previous
history. For some distance below the living leaves,
the trunk is connected by the withered, hanging,
but still persistent leaves of several successive
developments, a ragged drapery, of which we might
certainly say—

     " —— when unadorn'd, adorn'd the most."

The lower portion of the stem is, however, destitute of the decayed leaves themselves, the lozenge-
formed bases of them alone remaining, still separa-
ble, indeed, but sufficiently compact to make in the
aggregate a sub-cylindrical column of loose texture,
which may in familiar parlance be termed a *trunk*.
This portion is marked by alternate enlargements
and constrictions of the outline, which appear to
indicate seasonal growths.

    The specimen before us is about twenty feet in
height, exclusive of the crown; supposing these
swellings to mark a year's growth, and to be continued in the same proportion on that part of the

trunk which is masked by the decayed leaves as on the exposed part, we should conclude this tree to be about thirty-five years old ; for there are about thirty-four such swellings, each of which contains about four hundred of the lozenge-shaped bases of the fallen leaves.*

Remember, however, that we are looking at the Grass-tree, not as it now appears on the sandy plains of Western Australia, in the nineteenth century, but as it came out of the hands of its Almighty Creator at some precise but unknown period of past time.

This White Lily, crowned with its cluster of nodding flowers, magnificently beautiful, each a fair emblem of the spotless purity of a noble virgin—if we remove the soil from its base, we shall find that the stem springs out of a fleshy bulb. This is covered with thick yellow scales, by taking away each of which in turn, we see that the bulb is made up of such, surrounding the central mass which has pushed upward, in the form of the stalk, with its leaves and flowers.

---

* My observations rest on the fine specimen of this plant preserved in the British Museum. Dr. Harvey, however, says, " The growth of the trunk in *Kingia* is very slow, and a specimen about ten feet high may probably be some hundreds of years old." Report of Dubl. Univ. Zool. and Bot. Assoc. for Feb. 25, 1857. See the note *infra* on page 188,

Now the whole of this beautiful array which we see was formed last summer, when, if we had divided the bulb longitudinally, we should have

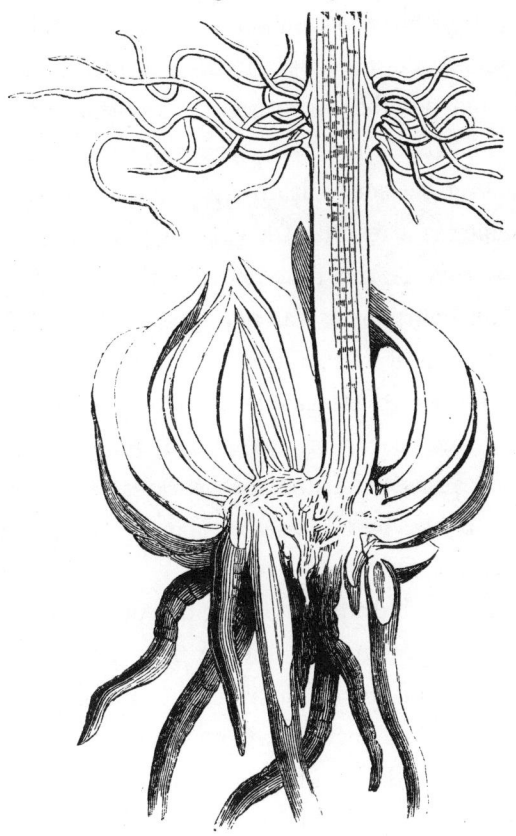

SECTION OF LILY-BULB IN JULY.

seen every leaf, every tiny blossom, folded together, and most snugly packed within the encircling scales, which are, indeed, undeveloped leaves; while from the base of the bulb so formed we should have seen pushed up on the outside of it, but yet within the common envelope of the exterior scales, the flower-stem of last season. There could not possibly have been this raceme of virgin blossom, if it had not been formed during the preceding season within the bulb; so that its existence is a record of a year's growth at least.

Yet this is the first hour of the lovely Lily's life; an hour ago it was not.

The face of the rugged cliff that rises perpendicularly above us was, a few moments ago, quite naked and bare, or diversified only by a few stunted prickly shrubs that sprang from its crevices. Now, by the mighty fiat of God, it is in an instant festooned from top to bottom with a most graceful drapery of round pale-green leaves, and slender stems no thicker than whipcord, and multitudes of spiral tendrils that climb, and hook, and catch, and entwine among the thorny bushes, and around the angles and prominences of the rock. We trace this curtain of verdure downwards, and find that it resolves itself into some half a dozen of wiry stems, that issue from different points of the surface of what seems a

boulder of brown stone, or a block of rough-hewn
timber, at the foot of the cliff.

TESTUDINARIA.

This angular block is, however, worthy of closer
examination. It is of no definite form, huge and
uncouth, lying as if cast accidentally on the
ground. Its whole surface is divided into a multitude of polyhedral pieces, that look as if they

had been cut into these forms by human art. Each division has a small angular face, and its sides display close parallel lines, all following the directions and angles of the outer face, but each line enclosing a slightly wider area than the one above it. These woody plates closely resemble in their angular forms and their concentric lines the plates of a Tortoise's shell, and hence our botanical friend, to whom we will appeal for an opinion as to the age of the block, will call the generic name *Testudinaria.*

"Well, I cannot give you any very precise judgment on the matter. The block itself is the tuber of a sort of yam, which grows above ground instead of below. It is a woody mass of great age. The angular plates are the bark, and they are so divided in consequence of the gradual growth of the tuber, tearing open its periphery to obtain more room. The concentric lines on the edges of the plates will not give us any adequate idea of the age of the mass, for though they indicate seasonal growths, the earlier layers have been worn away in the lapse of ages, and there are many layers of bark that have not yet been burst by the expansive force of the growing wood. It is known that these blocks are of very slow growth ; in tropical regions they last, with scarcely perceptible increase, from generation to generation. From such vague data as we possess, I might loosely conjecture this tuber to be a thousand years old."

We thank our scientific friend, and think it a very satisfactory report on an organism, which we saw called into existence five minutes ago, before our eyes.

Come away; for I wish you to look at this *Encephalartos.* A horrid plant it is, a sort of caricature of the elegant Palms, somewhat as if a founder had essayed a cocoa-nut tree in cast iron. Out of the thick, rough, stiff stem spring a dozen of arching fronds, beset with sharp, sword-shaped leaflets, but having the rigidity of horn, of a greyish hue, all harsh and repulsive to excess. In the midst of this rigid coronal sits the fruit, like an immense pine-cone.

The swelling column that constitutes the stem is but a mass of pith, surrounded by a thin case of wood, and enclosed by the remains of former leaves. The whole surface is covered with the lozenge-shaped scars of these, in vast number. Thousands of these there must be in this trunk of eight feet high, and a foot thick. The leaves

of the existing crown are few and very durable,
so that it would be no unreasonable conjecture to
suppose that this great Cycadaceous plant is seven
or eight centuries old.

ENCEPHALARTOS.

Nay, for this also has been created even now!
What shall we. say to *this* singular phenomenon?
Observe yonder gigantic Fig (*Ficus Australis*)
growing out of the face of that vast rocky pre-·
cipice. It is not so much to the massive grandeur

of the trunk, nor to the wide-spread head of dense foliage, that I call your attention, as to the broad expanse of roots, from the thickness of your body to that of your little finger, which have crossed and interlaced and separated and re-united, in all imaginable ways, until the whole forms a great flat network of wood, investing the surface of the rock, and following all its projections and angles with singular faithfulness, for a space of many square yards.

Would you not say, admitting that the Figs are rapid growers, that many years must have elapsed since the minute seed was dropped in yonder crevice, by some vagrant parrot that wiped his beak after breakfast on the point of rock? Would you not say that many years must have passed from the time when the tiny shoot peeped from the rocky chink, to the present moment, when the leafy honours of the crown above and the woody wall of the roots below combine to repay the protection which the plant in infancy received from its stony foster mother?

Of course you would; and most truly too, did you not know that the Fig-tree is now rejoicing in the first hour of its new-created being.

So with its noble congener here, the manytrunked Banyan (*Ficus Indica*). Although not an old tree, its canopy of broad downy leaves is already supported by so many secondary trunks, that it is not easy to say which of the larger stems is the mother trunk, and which the hopeful daughters. Every one of these stems, some just protruding from the horizontal limbs, others hanging midway between the leafy roof and the earth, some just inserting their slender spongy tips into the soil, others thick and pillar-like—is an evidence of progressive development, and therefore of lapsed time; only for the qualifying fact, that the development in this case is *prochronic*.

Here is the great *Euphorbia grandidens* of Africa. Its stout trunk is marked with a number of holes, some four or five inches apart, arranged in perpendicular rows. In some cases they are rather depressions or pittings than holes, and look like what would result from borings made with an auger in pitch in warm weather, the margins of which had nearly closed subsequently. What is the explanation of these marks? They are all records of time. From each of these spots once grew one of those angular prickly branches, that look like our commonest sorts of *Cactus*, and which are now confined to the summit of the trunk, arching out from it, somewhat like the branches of a candlestick.

It is the habit of this plant, when the stem has acquired a certain thickness, that the branches should, after a time, decay and drop off at the point of their union with the trunk, or rather a little below the surface, so as to leave the shallow holes or pits which we see. After their decadence, the growing bark gradually swells around the scars, and has a tendency to obliterate them. This may account for the non-appearance of them on the lower parts of the stem.

Here, then, are demonstrations of several successive stages of development. First, the stem must have been in existence before any lateral branches could have sprung from it. Secondly, the branch shot out. Thirdly, it put forth its spines and leaves. Fourthly, it died and sloughed away. Fifthly, the growing bark encroached on, and finally obliterated the cicatrice.

In this individual, all these stages are illusory, or rather they are prochronic.

See this noble Tulip-tree (*Liriodendron tulipiferum*), a giant of this primeval forest ; its towering trunk is crowned with a head of large massy foliage, of a rich deep verdure, among which shine numbers of great golden tulip-like blossoms, as fragrant as beautiful.

It is, however, the leaves that grow on the terminal twigs that I wish you specially to notice. These, which, as you see, are large, and of a remarkably elegant form, are fixed at the end of long petioles, which are set alternately on the twig. Notice, now, the manner of their development; the young unexpanded leaves grow within two large leaf-like bracts, forming an oval sac, which, as the young leaf increases, swell, and at length burst, and are left on each side of the base of the leaf-stalk. There is a succession of these. On this growing twig, for instance, I find three leaves already expanded (*a a a* in the accompanying figure), and as many pairs of these bracts (*b b b*) at their bases ; the twig is terminated by a pair (*c*) convex outwardly, and whose edges are in contact with each other; if, now, I cut off one of these (as represented at *d*), I expose the next leaf (*e*) folded together, and bent downward, in its pretty manner of *vernation ;* beside it is another pair of bracts (*f*), whose edges are not only in contact, but mutually adherent, and that with considerable force. On tearing these apart, I discover another smaller leaf, and another smaller pair of adhering bracts, which a gain contain a similar set, only yet more minute, and so on in succession, till I can no longer trace them.

TWIG OF TULIP-TREE.

Now it is manifest that the uppermost of the three leaves, together with the developing terminal bud, was at one time enclosed in the pair of bracts immediately below its base; that, before that, the middle leaf, with all above it, was similarly incarcerated in its own proper bracts; and, at a period anterior to that, the lowest leaf also. Each pair of bracts is therefore a record of a past period; and together they testify to a succession of past periods.

And yet their combined testimony is utterly worthless, because the noble tree was created in its magnificence this very day.

The beautiful twiner (*Bignonia*), which has cast its ample festoons over the topmost branches of yonder towering Mora-tree, almost concealing the natural foliage with its own elegantly pinnate leaves, and adorning it with its gorgeous trumpet-shaped flowers, is distinguished by a curious property, indicative of the years that have passed over it. In its adult maturity, as we now see it—the glory of this tropical forest—we should find, if we cut across the main stem, that its wood is divided into lobes arranged in a radiate or star-like fashion, like the divisions seen on dividing an orange transversely; and these lobes are thirty-two in number.

But this condition has not existed through the life of the plant. The wood has always been lobed, but the number of the divisions has varied, and that in geometrical ratio. Before the present stage, the constituent lobes were sixteen, which became thirty-two by the sub-division of each.

In an earlier stage there were eight lobes, and, earlier still, four, which was the commencing number; the duplication having proceeded in each case by the fission of each of the existing lobes into two.*

Now though this phenomenon will afford us, on the data we at present possess, no insight into the age of the plant, considered as an actual chronological period, an examination of a transverse section would always determine which stage is then present, and, by consequence, how many previous stages have been passed through. And thus we obtain a distinct clue to the former history of the organism, though we cannot mark it off into months and years.

Yet the fact of creation stultifies all the conclusions that we might form from such premises; since it does, *ipso facto*, contradict every such thing as a previous history.

On this *Anona* there is an intruder more strictly parasitical; it is a *Loranthus*, with long, club-shaped, richly-coloured blossoms. The branches of the supporting tree—a nurse who feeds her foster-child on her own vital juices—are overspread for a large space with the shoots; which,

* Gaudichaud : Recherches Gén. sur l'Organographie, p. 129.

I

springing each from its own disk, appear like so
many distinct individuals, but are really all parts
of a single plant, springing from a single seed.
(For this curious fact we are indebted to the observations of Mr. Griffith, who has investigated the
singular history of these parasites.)

The ripe seeds firmly adhere to the substance
on which they are applied, by means of their
viscid envelope, which soon hardens into a transparent glue. In the course of two or three days,
the radicle curves towards its support, and, as soon
as it reaches it, becomes dilated and flattened. An
union is gradually formed between the woody
system of the parasite and that of the stock, after
which the former lives exclusively on the latter,
the fibres of the sucker-like root of the parasite
expanding on the wood of the support in the form
of a *paté d'oie.* Up to that time the parasite had
been nourished by its own albumen, which is now
exhausted. As soon as the young parasite has
acquired the height of one or two inches, when an
additional supply of nourishment is required, a
lateral shoot is sent out, which is, especially
towards the point, of a green colour. This at one,
or two, and subsequently at various points, adheres
to the support by means of sucker-like produc-

tions, which are precisely similar in structure and mode of attachment to the original seminal one. The fibres of the parasite never penetrate beyond their original attachment; in the adult the sucker-bearing shoots frequently run to a considerable distance, many plants being literally covered with parasites, all of which have originated from one and the same seed.*

YOUNG PLANT OF LORANTHUS.

In this case, again, how delusive would be any

* On the development of *Loranthus*, &c. Linn. Tr. xviii. p. 71, (*abridged*).

inference of actual lapse of time deduced from the condition of a plant, which had been created as an adult capable of reproducing its race!

Here is a great impenetrable thicket of Prickly Pear. The delicate sulphur-hued flowers expand their broad bosoms to the sun, and the swelling fruit beneath is already putting on its lovely blush of crimson. How curious are the leafless but leaf-like dilatations of the stem—these flat oval plates of parenchyma, studded with clusters of woody and most acute spines! Every one of these expansions is an expression of time, as they are of course successive, though several may be formed in a single season; and not only so, but the tufts of spines, which grow at the points of intersection of crossing lines, in a network pattern, are all successive, appearing in turn as the expanded joint of the stem grows out.

The jointed dilatations themselves are, however, transitory; in the slow lapse of years the common woody axis enlarges, and the interspaces between the oval plates become gradually filled up with cellular tissue, and thus are obliterated; the stem, as may be seen in the central part of this spreading thicket, becoming round, almost smooth, and of dense woody texture. " This condition is the

result of many years," you say. It is so, in the
ordinary course of nature; but in the case before
us, it has been educed in a totally different
manner, and by a totally different energy, viz.
prochronically, by the omnipotent fiat of the
Creator.

We have emerged from the forest glooms, and
are come within the light and the music of the
sparkling sea. And here at its margin, washed by
its wavelets, there has been suddenly created a
Mangrove tree (*Rhizophora*), destined to be, doubtless, the fruitful parent of a grove, which by and
by will fringe this flat and muddy shore for miles,
shutting out the light and air which now freely
play over the beach, and keeping in, beneath a
long canopy of dense and leathery foliage, the
murky vapours which will rise from the decomposition of its successive exuviations.

As yet it is a single tree, but in its perfection of
maturity. And see how characteristically we find
here that singular structure, or rather habit, which
in Mangroves of normal development would be
the effect of age. The trunk springs from the
union of a number of slender arches, each forming
the quadrant of a circle, whose extremities penetrate into the muddy soil. These are the roots of

the tree—there are no others—that shoot out in this arched form from the base, or " crown" of the stem, taking a very regular curve of six feet or more in length before they dip into the mud. The larger arches send out secondary shoots from their sides, which take the same curved form, but in a direction at right angles to the former; and thus a complex array of vaulted lines is formed, which, to the crabs that run beneath—if they were only able to institute the comparison, must be like the roof-groins of some Gothic church, supposing the inter-spaces to be open to the sky.

Now, normally, it would require a lapse of several years from the first dip of the radicle of the seed into the soft soil, to form these arches, and to lift the axis of the tree a foot or eighteen inches above the surface. But here the same result is achieved in a moment, by the exercise of crea-tive power.

Look at this *Eriodendron*. What a magnificent accumulation of vegetable cells is here! Its colossal trunk rises in naked majesty, a massive column, to the height of a hundred feet, without a branch. And then what branches! Those limbs them-selves are of the bulk of ordinary forest trees; they break out, three or four on the same plane, and radiate horizontally to a vast distance, supporting a noble flat "roof of inwoven shade."

SILK-COTTON TREE.

Perhaps the most remarkable feature of this majestic tree is found at the foot of the trunk,

which sends out vast spurs, radiating in all directions, and extending to a circle of seventy or
eighty feet in diameter.   These spurs take the
form of perpendicular walls of timber, commonly
not more than six or eight inches thick, pretty
equal in their thickness throughout, and varying
in height from fifteen or twenty feet, where they
spring from the trunk, to the point where they
enter the soil.

Now the Silk-cotton tree has not had this form
through its life.   When young, say up to twenty
or thirty years old, there was no appearance of
spurs; the trunk was covered with a green bark,
and was studded with great triangular low spines,
an inch in diameter.   And, what had a curious
effect, the middle of the stem swelled into an ovate
form, quite symmetrical on all sides. But, as years
passed, the ventricose form of the trunk was gradually lost; the bark became of a hoary grey hue
or even almost white; the three-sided prickles disappeared from the bole, and were retained only on
the upper surfaces of the limbs; and the great
lateral buttresses began to fill up the angles which
had hitherto existed between the trunk and the
main horizontal and superficial roots.

I called the noble tree before us an accumulation

of vegetable cells.   And viewed in that aspect, what an irresistible evidence of the lapse of time does this vast organism present to us! since the whole of this immense structure originated in a single cell, which, by repeated acts of self-division* (or, possibly, other modes of reproduction), has gradually built up the mass.

Yet such a retrospect would be most fallacious in the case before us, since the plant, as a perfect compound organism, with its parts—root, trunk, limbs and leaves, and its tissues—cellular, fibrous, and vascular, has been produced by the instantaneous putting forth of the Divine volition.

Once again.   More gigantic even than the towering Ceiba, this immense Locust-tree (*Hymenæa*) appears to penetrate the very sky with its crown of foliage, which is so remote from the earth, that our eyes cannot avail to discern the forms of the leaves.   The straight columnar trunk, like some triumphal monument in the midst of a

* "Each and every plant is at first a cell."—"New cells can never be formed externally to, but only within, other cells already formed."   (A. Braun, on the Veg. Indiv.)

"The process of the propagation of cells, by the formation of new cells in their interior, is an universal law in the vegetable kingdom." (Schleiden; Grundzüge).

"Cell-formation in plants takes place only in the cavities of older cells." (Mohl, on the Veg. Cell.)

I 3

great metropolis, is of so vast a bulk that a dozen of such men as you and I could scarcely embrace it with stretched arms and joined hands.*

Can our friend, the vegetable physiologist, help us here to form a notion of the time which would be required for the production of this tree in the ordinary way? It is the last favour we will ask of him to-day. Come, Sir, give us your thoughts on the matter.

The Botanist.—" There is a principle which, in trees of this character, namely, such as are of exogenous structure, will determine the age with very close accuracy. Each generation of leaves sends down woody fibres, which unite into a cylinder on the outside of the wood previously formed, and beneath the bark.

" Now, as these cylinders are in general sufficiently distinct, in those trees which renew their leaves but once in a year, it will be enough to count the concentric circles which appear on a transverse section of the trunk, and we shall obtain the number of years during which the tree has existed. In the case of this great Locust, the rule, to be sure, is rather difficult of application in that way ; a transverse section of this trunk would cost

* See Von Martius, on the Brazilian Locusts.

a little labour. But with this circular saw, which I always carry about with me for investigations of this sort, I can take out a horizontal cylinder on each of two or three sides of the tree, by counting the layers in which I can form a tolerably accurate estimate of the number in the whole diameter.

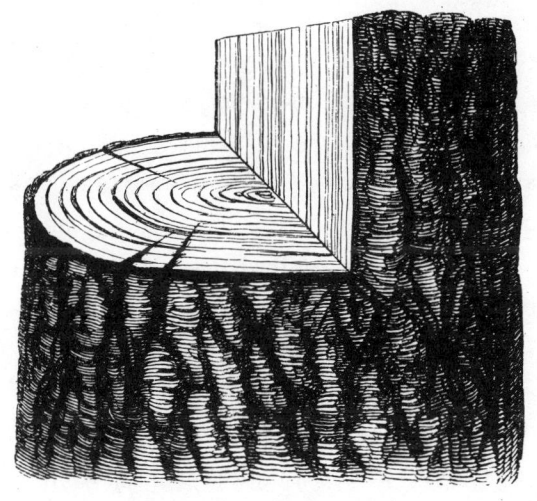

SECTION OF EXOGENOUS TREE.

" See; in these cylinders, which do not materially differ, there are seventy-two layers in a foot, that is, each layer is one-sixth of an inch wide. The trunk is, at the part I have tested, about fifty feet in diameter, or twenty-five feet in radius; which would therefore contain just eighteen hundred such layers. As the deposition of new wood, however, is generally more abundant in youth and middle life than in age, the layers are probably a little wider, that is, fewer in a given space, as we approach the centre. For this we must make allowance, and may conjecture that this tree is probably not less than one thousand five hundred years old."

Now whether the premises of the botanist will bear out this conclusion or not, is not a vital question. For the question at issue is, not, *How long* it has lived, but, *Whether it has lived at all*, before the present moment. It is enough for our point that the tree does, in its concentric zones, afford ocular evidence of successive epochs of growth. And the proof of this would be equally good, if ten layers were deposited in a year, or if one deposit were made every ten years; equally good, if there were fifteen hundred zones, or if there were but five. It would be easy to confirm the testimony of the zones by that of other parts of the structure. The dimensions of the tree itself bear a fixed and, to a certain extent, recognisable ratio to its age; every leaf on a given twig has been successively developed from a leaf-bud, the opening of which and its elongation into a twig occupied, normally, a definite period; each bough, each of those mighty limbs, was once a twig, was once an undeveloped leaf-bud, whose expansion to its present condition was a process, of which time was an inseparable and, within certain limits, a mensurable element.

If, then, we were precluded from examining any other organism, as it proceeded from the formative hand of its Creator, than this single tree, we should be amply warranted in inferring a past existence (be it longer or shorter, which is no matter) from the phenomena of its structure, which inference the fact of its creation would flatly contradict.

VIII.

PARALLELS AND PRECEDENTS.

(*Invertebrate Animals.*)

" There is a kind of character in thy life
That to th' observer doth thy history
Fully unfold.——"   (*Shakspeare.*)

LEAVING the vegetable kingdom, those organisms
which, though beautiful indeed and instructive, are
yet inanimate, let us seek others which are endowed with a higher style of life, a life which is
distinguished by a measure of consciousness of
the exterior world, and a perception of relations to
it.   Let us look for animals.

We retrace our steps to the verge of the rippling
sea, where the belt of umbrageous Mangroves
fringes its margin.   Beneath the arching roots of
these are now reposing in the warm sunlit shallows
many creatures which number this as the first day
of their existence.   It is their natal, or rather (to
make a word) their *creatal* day.

Here is a specimen of the Sea-pen (*Pennatula*),
closely resembling a rather thick and fleshy feather,

with its quill-end inserted in the tenacious marl which constitutes the floor of the sea along this shore, and with the greater part of its body, including all the pinnated portion, erect, and waving lightly in the gentle swell of the bay. Its central stem is beset on each side with about twenty-five horizontal purple pinnæ, and each pinna bears from five to fifteen polypes with eight tentacles each.

Let us wade out to yonder reef. See this great mass of Millepore, growing in thin irregular perpendicular plates, which join each other at various angles, so as to form a large open honeycomb-like structure, much resembling the second stomach of an ox. It is covered with what appears a thin stratum of fawn-coloured jelly, but this consists of innumerable disks, which protrude from myriads of orifices not larger than those produced by the punctures of a fine needle; as we may discern by touching the soft slimy surface, when the whole retires, and leaves apparent only the white stony surface dotted with numberless holes, within which the disks have disappeared, and whence they will again presently re-appear.

Here too is a massive shrub of stone, a noble example of the Muricated Madrepore. It consists of a great multitude of short branches, which are themselves branched and branched again, every part covered with little mammillary warts, and pierced with innumerable holes in which stand radiating plates of the common stone. Out of these plated orifices, especially those towards the tips of the branches, for the older ones are empty and dead, we see perpetually peeping forth, expanding for an instant, and then coyly withdrawing, lovely little green disks, surrounded with thread-like tentacles; and from the extreme end of each branch there protrudes one exactly similar to the rest in all respects, except that it is nearly twice as large. Here then are the living architects; these have secreted within their gelatinous membranes the calcareous atoms, whose aggregate forms the stony shrub before us.

Shall we try to estimate the number of polypes that have been occupied in building this tree? There are about a hundred branches, which, taken one with another, and followed along the sinuous course of their many branchlets, we may estimate to average a continuous length of eight feet each; that is, 800 feet of branch in all. Now we may consider these branches as averaging a thickness of two inches and a half in circumference, which gives us a surface of 24,000 square inches. Finally, there are about ten polype-cells in each square inch; and thus there are or have been in this coral-mass, nearly a quarter of a million of polype inhabitants.

MURICATED MADREPORE.

But look at this dark crimson edifice of many stories, tier above tier, each horizontal floor of red stone sustained by a multitude of slender cylindrical pillars. When we look closely at them, we see that the pillars are tubes, perforating one or more of the floors, from the lowest tier to the uppermost.

Have we any clue to the age of these corals, or to that of either of them, supposing we did not know that they have been created to-day? Not definitely, perhaps; but indefinitely we have, certainly. In the case of the Sea-pen, the polypes have all been formed in succession; as also in that of the stony Millepore and Madrepore, with this addition, that every newly formed polype deposited an increase to the stony substance, which thus went on increasing till the great foliated or ramified mass that we see was formed.* And so, with this series of floors and pillars, which is the solid portion of another coral-polype, the Organ-pipe (*Tubipora musica*).

Every one of these stories has been formed in succession. From the tips of some of the tubes we see protruding an elegant polype of an emerald-green hue, having eight starry tentacles, and giving

---

* The origin of coral-stocks is minutely described by Ehrenberg, in the Abhandl. for 1832, where he makes the following remarks:—"The coral mass is neither a mere structure composed of many animals arbitrarily conjoined, as Ellis supposed; nor one single animal with many heads, or with simple furcations, as Cavolini maintained; nor a vegetable stem with animal flowers, as Linnæus expressed it; it is a body of families, a *living* tree of consanguinity; the single animals belonging to it, and continually developing *upon the primary ancestor*, are entirely isolated within themselves, and capable of complete independence, *although unable to achieve it.*"

off from its base an enveloping membrane, which
spreads over the rim of the tube and descends on
the outside to the floor. By means of this vascular
membrane, both tube and floor have been formed.
Calcareous particles, deposited, one by one, in its
substance, gradually built up the tube of the

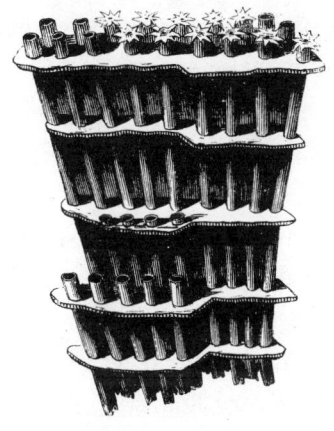

ORGAN-PIPE.

primary polype, or probably the tubes of the first
series, the basement or ground-floor. When these
tubes had arrived at a certain height, all simultaneously began to develope the fleshy membrane
horizontally, which expanded until that from each
touched that from its neighbour, with which it
united. Meanwhile the calcareous deposition went

on in this horizontal layer, and thus the first floor was made.

Now from the living vascular upper surface of this layer sprang up at certain spots buds,* off-shoots of the common flesh, which soon rose into columns, and, by a process of calcareous deposition, became tubes with terminal polypes, which in turn spread out a horizontal layer, and thus the second floor was built. Hence a new race of polypes budded, which by and by formed the third floor; and so on in succession until the series had attained the height which we see.

If we assume one of these stories to be the growth of a year,† we have ocular evidence in this specimen of six years' age, for here are six successive floors. But no: for it was created complete, as we see it, this very hour.

Yonder goes a *Medusa*, pumping its way laboriously, yet not ineffectively, just beneath the

* This is not quite in accord with Lamouroux's account; but it is more consistent with what we know of polype-growth.

† We lack precise data on which to found conclusions as to the actual rate of growth of many animals. Sir John Dalyell's famous Actinia, now in the possession of Dr. Fleming, affords us a proof that the Zoophytes are long-lived, and slow in attaining maturity. It will be readily seen, however, that the argument in the text does not depend on the actual period evolved. The lapse of *a* period of time, no matter how long, is the only essential point.

surface of the clear wave. It is a great affair, nearly a foot in diameter. Have we, from merely examining its appearance and structure, any criterion by which we can guess whether it has lived an hour, or a year, or ten years? Surely we have; for this mass of clear jelly is composed, like all other organic bodies, of cells, which have been gradually generated, by nutrition and assimilation, from the embryo.* This process must have occupied many months, if not several years; but the history of this Medusa did not begin when it took its present umbrella-like form. Shall we trace it back a little farther?

At some time back, then, this creature detached itself as the terminal one of many little saucer-like bodies, which had been for some time previously forming by the gradual constriction of a thick fleshy stem. Before the constriction began to be visible, this stem was the body of a white Hydraform polype, affixed by its base, and furnished at its free extremity with thirty-two tentacles. It had lived several years in this form, developing many Hydroid polypes, just like itself,

* "All the component cells of any one organism may be considered as the descendants of the primordial cell in which it originated." (*Dr. Carpenter;* Comp. Physiol.; p. 396. 4th Ed.)

by successive gemmations. Before it took this shape, which it assumed gradually, its tentacles being developed in geometrical progression, 32 from 16, from 8, from 4,—it was a soft ovoid planule clothed with vibratile cilia, which swam freely in the sea, like an *Infusorium*.

Thus the physiologist would confidently assign to this Medusa an existence of several years, as an independent organism; *nor could his conclusions be controverted*, except by the knowledge of the fact that the Medusa *has been but just now created.*

We pass on. Here is an *Echinus.* Let it be borne in mind still, that we have, *in idea*, the power of pursuing our researches on each creature at the moment which follows that of its creation; and that, when that actually was is of no consequence to our investigation.

Here then is this new-made *Echinus sphœra*, a somewhat conical globe of three inches diameter, which is covered with a forest of spines, pedicellariæ, and suckers, and which glides majestically along, with an even but slow progress, over rock and reef. Its vitals are enclosed in a hollow box of calcareous shell, which is built up of nearly a thousand pieces. This specimen, which is rather below than above the average size, is formed of ten meridional rows of large plates (the inter-ambulacral), and ten of small (the ambulacral). The former series are each composed of thirty-two plates, making in all three hundred and twenty; the latter have just double that number, making six hundred and forty; thus this Urchin's box is built up of nine hundred and sixty plates; every one of which is of definite shape and angle, and fits into the angles of its fellows with the accuracy of the most skilfully constructed cabinet-work.

Now every one of these plates is an eloquent witness to the past life-history of the Sea-urchin. For the reason why the enclosing box is made of so many pieces is, that it might gradually expand and enlarge its capacity with the ever increasing requirements of the soft organs within. Every plate is enveloped by a vascular flesh, from which the calcareous particles are deposited in a constant and perfectly uniform ratio; and thus all the constituent plates are continually enlarged by additions to both the internal and external surfaces (increasing their strength), and to their sutural margins (increasing their combined capacity), until the adult dimensions are attained.

The size of the new-born Echinus is not nearly equal to that of one of these plates, and the progressive increase of the plates by deposition on their edges has certainly taken several years to accomplish.*

The same result is inferrible from the structure of the spines with which every plate is armed. Each of these is a very long cone of calcareous matter, arranged in minute oval chambers, divided by thin glassy walls, and deposited particle by particle from the thin stratum of living flesh with which each has been invested from its first embryonic development.

But of this *Echinus*, as of the *Medusa* before, we find a history anterior to either box or spines. Its first appearance in this stage of existence was as a barely-visible circular disk, constructed on the outside of the stomach of a singular transparent organism, much like a Medusa, but of a domular form with four or six legs, stiffened by calcareous rods, and a crowning pinnacle. For some undefined time this gelatinous dome had been gliding with a stately movement through the open sea, before there was the least trace of the disk, which afterwards grew to the *Echinus*. In its earliest

---

* I conclude so; because I have kept specimens of *Echinus*, not full grown, in healthy condition, for nearly a year, without any perceptible increase in their dimensions.

condition the dome itself was a soft, spherical, mulberry-like *Infusorium*, covered with vibratile cilia; this altered its form to that of a three-sided pyramid, and this to the vaulted dome.

Clearly, therefore, we have a right to infer a past history of the Urchin, and that of not a few distinct stages. But no; the specimen has commenced its history within an hour!

Yonder Feather-star (*Comatula*) notice; which, having just now started into mature life at the almighty fiat of its Creator, goes careering joyously through the sea, expanding and contracting its many-jointed and feathery arms, as if it had been accustomed to the alternation for a long life, and ever and anon settling itself by grasping the points of rock with its dorsal claws. You would hardly think that those flexible and slender arms were made of stone: yet they are; every joint of the stems and of their pinnæ is a vertebra of stone (precious stones, you will say—topaz and ruby—from their brilliant hues), which has been formed and deposited atom by atom, by the slow and gradual process of secretion of calcareous matter; the lime having been primarily collected from the sea-water which held it in solution. At least, such is the physiological deduction.

K

But there was a period in the *Comatula's* history when it was not a free-swimming star, but a lily-

COMATULA AND YOUNG.

like flower of ten slender fringed petals, seated at the summit of a long stalk, with a central columnar **axis** of stone. Before that, the flower-head had a bud-like figure, and the petals were minute and destitute of lateral fringes; and earlier still, it was a tiny gelatinous club without any development of stone, affixed by a spreading base, and shooting forth from the top a few pellucid processes. Earlier still, it was, no doubt, an infusory-like gemmule, clothed with cilia.

Through all these successive stages, which, of course, occupied a considerable period of time, we should certainly affirm the Feather-star to have passed, did we not know that it has this very hour burst into existence.

That Panther, whose tawny fur studded with black rosettes appeared so beautiful as he bounded with agile grace from glade to glade just as we emerged from the forest, contains within his intestines, though you cannot see it, a mature Tapeworm. The body of this parasite consists of some hundreds of square flattened segments, each of which includes a complicated generative apparatus, equal to the production of thousands of fertile ova. Is not this an evidence of age? For, first of all, consider that the formation of each of these hundreds of joints has been a work of development from the anterior parts; and therefore they record as many distinct and successive processes as there are segments. And, secondly, remember that the *Tœnia* did not commence existence as a *Tœnia*, nor in the conditions in which it now exists, within the bowels of the Panther. It looks back to another form, and to another living *nidus*.

There was a time when this parasitic creature had no ribbon-like body of flattened generative segments. There was, indeed, the same curious head, a tiny globose knob at the extremity of a slender neck, furnished with the same array as now, of rows of hooks and sucking disks. But in place of the segments, the neck merged into a membranous bladder distended with clear fluid. It was not a *Tœnia* then, but a *Cysticercus*.

Its home was at that time the interior of a living animal on whose vitalized juices it was sustained, but that animal was widely different from its present patron. It was an Antelope, that cropped the wiry grass and aromatic shrubs of the arid plain.

Earlier still, the germ of this *Tœnia* was an egg lying on the ground, having been discharged from the rectum of another Panther, in the bowels of which it had been developed by one of the segments of a former *Tœnia*.

Let us now trace the history of this organism

onwards from the point at which we have arrived in our retrograde researches.

The parent *Tænia*, still snugly ensconced in its obscene abode, partially matured and then separated the ultimate generative segment, containing many thousands of ova, far advanced towards perfection. The detached segment now became enclosed in the fæces of the Carnivore, and was at length discharged, enveloped in the pellet. The eggs, acquiring maturity, were hatched, and the infant worms individually scattered themselves among the surrounding herbage.*

One of these was devoured with the herbage by a grazing Antelope, and having safely escaped the perilous ordeals of mastication and rumination, passed into the stomach of that Ruminant, whence it soon made its way by some unknown but unerring route to the liver, in the parenchyma of which organ it rapidly developed the cyst, which gave to the present stage its proper character.

The Antelope fell a prey to the ferocious Cat; its flesh was quickly digested in the stomach, but

---

* I am not aware that this stage of the Entozoon has been actually observed; but from what we know of its previous and subsequent history, the correctness of the statement in the text will scarcely be disputed. (See Prof. Owen: Comp. Anat. of Inverteb. Ed. 2. p. 74.)

the gastric juice produced no effect on the *Cysti-cercus*. This parasite had merely changed its residence for one more commodious, or at least more suitable for its further development. It presently attached itself to the walls of the intestine by means of its oral hooks and suckers, and, getting rid of its vesicular sac, with its fluid contents, probably by absorption, it began to develop, joint by joint, that immense ribbon, which it possesses now, and which constitutes it a Tapeworm.

Such is the " strange eventful history " of this repulsive creature ; a history legitimately deducible, in all its stages, from its presently-existing condition. But it is a history altogether illusory. The *Tœnia* never was a *Cysticercus :* the Panther is as yet guiltless of capricide : it is this moment called into being, and the Tapeworm begins existence within it.

This lump of red sandstone that has been rolled about in the sea, till all its points and angles are worn smooth, is now roughened again by the close and firm adhesion of extraneous substance, in the form of a cluster of shelly pipes, which twine irregularly over the surface of the boulder, and then start up erect with open mouths. These are the tubes of a species of *Serpula,* and the worm itself is seen now slowly emerging from one of them, and introducing its conical stopper, and elegant fans of white and scarlet filaments, to the genial daylight.

Observe, however, that the tubes are not of the same diameter throughout. At the point where they start up from contact with the stone, they are considerably smaller than at the tip; and if we trace back the adherent portion along its tortuous course, we find that it constantly diminishes until it is but a slender white thread of stone. Now this slender extremity was formed first; and as the worm itself grew, so it progressively required a larger and yet a larger habitation; which was readily provided of the due dimensions, because the material, which is limestone, was secreted by the swollen collar of the worm, and being freely poured out as required, was moulded of the proper calibre by the rotatory motion of the animal, combined with the special use of certain tactile organs for the purpose.

The shelly tubes themselves afford us ocular evidence not only of their progressive formation, but also of the successive steps by which this was effected. For at certain intervals of their length we perceive rings of the common stony substance, which mark the rim or mouth of the tube as it existed after each periodic increase. The mouth of the tube is, as we see, slightly expanded in a trumpet fashion; but as the general cylindrical

SERPULA.

figure is to be maintained, the next deposit of calcareous matter is not made at the very edge of the lip, but on a ring a little way within the margin, whence it is carried up, leaving the former margin slightly projecting.

Who could hesitate to assert that a history of

past time is legibly written in the annulations of these stony tubes? And yet the creatures, with their tubes, have been but this instant created.

But here is a tube of quite another construction, though inhabited by a kindred worm. It is wholly built up of sand, the inimitable architecture of the indwelling *Terebella*, who has thus succeeded in performing a task which defied the efforts of that too industrious artizan,—the familiar of the renowned Michael Scott.* Our worm has certainly spun a rope of sand, and one which holds together with surprising tenacity.

The instrument which our little architect wrought with are the long tentacles, which, like a tangled tuft of yellow sewing-cotton, twist and twine over the floors of sandy pools. Nothing at first sight seems less adequate for the purpose than those very slender, soft, and flexible threads. Dr. Williams shall tell us how they are used. "They consist of hollow flattened tubular filaments, furnished with strong muscular parietes. The band may be rolled longitudinally into a cylindrical form, so as to inclose a hollow cylindrical space, if the two edges of the band meet; or a semi-cylindrical space, if they only imperfectly meet. This

* See Notes to "Marmion."

K 3

inimitable mechanism enables each filament to take up and firmly grasp, *at any point of its length*, a molecule of sand; or, if placed in a linear series, *a row* of molecules. But so perfect is the disposition of the muscular fibres at the extreme free end of each filament, that it is gifted with the twofold power of acting on the sucking and on the muscular principle. When the tentacle is about to seize an object, the extremity is drawn in, in consequence of the sudden reflux of fluid in the hollow interior; by this movement a cup-shaped cavity is formed, in which the object is securely held by atmospheric pressure ; this power is, however, immediately aided by the contraction of the circular muscular fibres. Such, then, are the marvellous instruments by which these peaceful worms construct their habitations."*

Since the slender tentacles are the implements by which the sand-tube is thus built up, it is manifest that the existence of the tube must be subsequent to the existence of the tentacles. But the *Terebella* was at one time without tentacles; so that its history certainly reaches back to a date anterior to the existence of a tube. Several stages of life have intervened between that distinguished

* Report on Brit. Annelida, p. 194.

by the present worm-form, and its infant condition, when it swam as a ciliated undivided monad.

So, at least, we conclude from physiological data; but our conclusions are false, because contradicted by the fact that the mature animal with its case has been just now created.

Let us forsake the ocean-shore, and walk again through the glades of the virgin forest. A White-ant (*Termes*) crosses our path, and, by tracking him home, we speedily discover his dwelling, an enormous structure composed of gnawed wood cemented with an animal secretion, and formed into thin but very firm and hard layers. Swarms of labourers are passing in and out; and, on our breaking away a portion of the edifice, out come crowding the warriors, with formidable jaws extended widely, ready for the fight. In the interior we find numerous chambers stored with food, and nurseries occupied by young and eggs, the number of which is every hour increasing by the oviposition of the gravid female,—the queen of the city—who is lodged in an apartment in the very centre of the whole.

The entire edifice has been built around her; she is the hope of the colony, the only mother in this vast assemblage. It is therefore through her that we must look for a past history; and in her we find it. Some months ago, when she was not more than one thousandth part as large as she is now, though then adult, she migrated from some other city not less populous than this is now. It was just before the periodical rains, when, at the time of the great annual swarming, myriads of winged males and females were evolved from the pupa state, and flew out from their native city. This individual female was found by some of the workers that now compose this colony, and was immediately selected to be at once their prisoner and their queen.

We thus trace our great egg-laying Termes to a city of last year's building, in which for a time she was in an immature condition as a nymph, and before that passed a still less-developed stage as a larva. Hence her life-history goes yet farther back to an egg, originally laid by a former female in exactly the same circumstances as those in which we find this guarded and immured individual.

Thus we reason; but the female, with her host of attendants, and the house, which is inseparable from their present stage of existence, has been created to-day.

See that creature which with loud ringing hum is whirling round and round the tassel-like blossoms of this noble *Eugenia*. You would think it a bird from its massive size, but it flashes and sparkles in the sun, like a great jewel. Now it suddenly alights on one of the crimson flowers, and you may perceive that it is a beetle ;—a beetle of vast size, and glittering like a lump of burnished metal;—it bears the name of Goliath,—a giant clad in polished armour.

This is his first hour of existence; now for the first time has his nervous system responded to the stimulus of the sweet air and genial sunshine. An hour ago he had no nervous system; no system of any sort; no life; no being; no anything;—he was not until this hour.

Yet if we were to ask a friend conversant with entomology his opinion on the age of this insect, he would immediately give it; not, however, as an opinion, for he would repudiate the uncertainty which such a word implies, but as an indubitable fact, resting on the infallible grounds of constant observation and undeviating experience.

" This fine *Goliathus*," he would say, " has not long, probably, emerged from a hollow case of oval form, made of particles of earth agglutinated

together by a secretion from the mouth of the
larva, and concealed under the surface of the
ground. Within that sepulchre it has left its
cerements,—the shrivelled skin of the pupa, in
which it had been wrapped up motionless like

GOLIATH BEETLE, AND PUPA CASE.

a mummy, for several weeks prior to its appearance as a glittering beetle. The construction of
the oval cell was the last act of the larva, a thick,
massy, heavy-bodied grub, which had fattened for
years by feeding on the roots of plants beneath the
soil. Four years passed away* while yon beetle

* We have no direct observations, that I am aware of, on the
larval state of the African *Goliathi;* but their near ally, the *Ceto-*

lay on its side, darkly labouring at this occupation; and before that it was a minute egg for some weeks. The specimen before us cannot be far short of five years old."

No such thing: the witness is at fault: the *Goliathus* is not *an hour* old.

Take notice of the swarm of Gnats, which, like a dim cloud, are uniting in choral dance and song in the beam of the setting sun. Every member of the band that "winds his shrill horn," has had an aquatic before he had an aërial existence. A week was spent, in lobster-shape, with two breathing tubes on the summit of his body, in passing alternately from the bottom to the top of yonder stagnant pool, and then back from the top to the bottom. And a month was occupied in pretty nearly the same employment, but in another mask,—in fish-like form, with the star-tipped breathing-tube projecting from the side of the tail. But for some months earlier still it was a little lenticular egg, which was agglutinated with a number of others into an oval concave boat, that floated to and fro on the surface of the pool.

And there was something worth observing in

nia aurata of Europe, passes four years in the grub condition, as does also the *Melolontha vulgaris,* another lamellicorn beetle. The *Lucanus cervus,* or Stag-beetle, continues a larva for six years.

that tiny skiff of eggs; for it did, in its artful construction, carry the evidence of time back to a former generation. The eggs individually and separately would have sunk to the bottom of the water; it was, however, essential to their life that they should be in contact with the air as well as with the water. Hence they were so arranged in the aggregate, that the mass should swim, though the constituent individuals could not. To effect this, the parent Gnat, resting on the calm surface of the pool, crossed her two hind legs, and laid an egg perpendicularly in the angle so made : others were added in succession, all maintaining the perpendicular position, all glued together by a cement that resists water, but so arranged, the crossed legs being still the mould, that the outline should be spindle-shaped, while the summits of the central eggs, being a little lower than those of the outer ones, gave a concavity to the boat. So buoyant was it when finished, and the mother's legs withdrawn, that even a drop of water falling full upon it from above, would have failed to submerge it. There it floated, week after week, and month after month, all through the winter, till the genial sun of spring hatched the fish-like larvæ to begin their wriggling existence beneath the surface.

Now may we not say with confidence, that the sounding-winged insect looks back to the pupa, the pupa to the larva, the larva to the egg-boat? And more, that the form of the boat,—a form so essential that it could not have lived without it,—looked back to the crossed feet of the mother-gnat, the impress of whose angle its extremities sustained?

Of course we might reason thus: but yet we should be at fault; for the ringing swarm of merry Gnats has been this very evening created.

The Case-flies (*Phryganea*) that look like delicate moths of sober-brown hue, flitting over the surface of the pond, have, like the Gnats, spent a considerable time under water. When they were larvæ, they industriously col-lected small shells, fragments of stone, bits of reed, and the like matters, and, connecting them together with strong silk, made out of them slender tubes, in which they sheltered their soft bodies from harm, while their hard polished heads and shoulders projected from the open end. And after having lived through the winter (at least, but I rather think

LARVA OF CASE-FLY.

more than *one* winter) in this state, each closed up the entrance of his castle, by spinning across its open end, a transverse screen of lattice-work, made of very strong and stout silk, which, while it should serve the purpose of keeping out evil-minded intruders, during the helpless inaction of the pupa, should at the same time admit the free ingress and egress of water necessary for its respiration.

The life of the larva, and the exercise of these, its curious instincts, are, together with the duration of the pupa stage, inseparable precedents of the imago state in which we now observe the flying insects. No, not "inseparable;" for in this case, at least, they had no existence in time; they are prochronic developments.

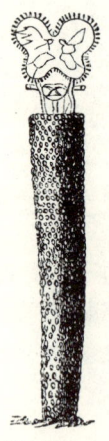

MELICERTA.

In this pond at our feet there is an object worthy of a moment's observation, minute though it is, for it is only visible as a speck to the unassisted eye. On one of the whorl-filaments of this tuft of *Myriophyllum*, there stands up a cylindrical tube, firmly adherent to the plant by its foot, but free at its upper end. Small as it is, this chimney is built up of hundreds of pellets, solid, round, and yellow; placed in sym-

metrical order, and firmly cemented together. What
has made this tube ? Ha! here is the little architect
ready to answer for himself; he thrusts out his head
and shoulders from his chimney-top, and announces
his scientific cognomen as *Melicerta ringens.*

Look! he is in the very act of building now.
Did you see him suddenly bow down his head and
lay a brick on the top of the last course? And
now he is busy making another brick ; his mould
is a tiny cup-shaped cavity just below his chin :
his material the floating floccose atoms of vegetable
refuse. Cilia along his flower-like face collect these
atoms into a stream, and pour them into the cup ;
and cilia within the cup whirl them rapidly round
and round in many rotations, until with the aid of
mucus they are somewhat consolidated into a round
pellet. The brick is made, and nothing remains
but that it be deposited next the former, in regular
progression, and this is done by the tiny τέκτων,
suddenly bending his head forward, and bringing
the chin-cup with exact precision to the spot.

And how long has he been engaged in this piece
of work? Little more than a day. It was commenced yesterday, when the creature was not more
than one-third as large as he is now. But he had
lived a few hours before the commencement of

his work. He was a rover before he began to be a house-keeper. In that early stage of youth and freedom, before he had made up his mind to settle in life, he had no chin-cup, no flower-like face, and of course no tube. A cylindrical gelatinous pellucid worm, he issued out of the egg, with a brush of cilia on his crown, and danced waywardly through the water. While thus occupied, his form underwent some preliminary modifications, and at length was sufficiently matured, to enable him to choose a spot for the passing of his future life, and to commence the building on which he is still engaged.

Not so. The pellet which he deposited when we began to look at him, was the first he had ever made ; he had been created but that moment ; and all the previous pellets of the case had been called into being just as we saw them. They were built up prochronically.

I tear a piece of bark from the trunk of this half-decayed tree, and have disclosed amidst the rank-smelling damp and rotten wood, a large *Julus,* a slow-moving creature, with some hundred-and-fifty little twinkling feet. As this specimen has attained its adult condition, it must be at least two years old ; for it does not acquire its reproductive organs and perfect development till that age.*

This creature has passed through a rather curious history of evolutions. The egg from which it was produced was lodged in a chamber excavated by the parent, a few inches below the surface of the rotten mould. From this egg proceeded a little kidney-shaped body, without limbs or motion, completely enveloped in a swathe of delicate transparent membrane. About a fortnight it remained in this helpless state, during which its organs had been forming out of the constituent cells, by repeated subdivision, and definite arrangement. At length it burst its cerement, and a minute Julus appeared, not more than $\frac{1}{200}$th of an inch in length, composed of a head with antennæ, and a body of eight segments, of which the first three carried each a pair of legs.

All the multitudinous limbs which we see in this adult have been produced in successive moultings, and all the numerous segments have been produced by the subdivision of the last but one,—that is the joint preceding the anal one,—six at a time.

By the time the little animal was ready for the second sloughing, that is, in about a week after

* Fabre; Ann. d. Sci. Nat.; iii. 1855.

the preceding, three more pairs of feet were seen, which had budded from the fourth, fifth, and sixth segments, but which were as yet closely packed down beneath the investing skin; the seventh segment also was obscurely marked into six divisions. The skin was now thrown off, and these changes were perfected; the little Julus now had six pairs of feet, and thirteen segments.

This process was repeated again and again; the new limbs always developing on the segments last produced, and six new segments being always formed out of the existing penultimate. And by this gradual succession of development, the animal has attained the number of limbs and segments which we now perceive. The antennæ and the eyes have likewise passed through successive stages.

We have a right to infer the lapse of a period sufficient to produce these changes, for we see their indubitable results; but our inference would only lead us astray, because we have not allowed for a disturbing influence,—that of the Law of Creation. This is the Julus's first hour of life.

See, on the trunk of that towering *Cedrela*, a round hole, out of which a large Beetle is in the act of emerging. It is a noble *Buprestis*, encased in glittering mail, of the most refulgent metallic

splendour, crimson, gold, and green. Can we find
any clue to his age? Yes: the white grub has
rioted and fattened in its burrows in the timber of
this tree for many years; ever gnawing away with
its horny auger-like jaws the solid wood in tortuous
galleries, which constantly enlarged, as it progressively grew, while its wake, as it advanced, was
partially filled by its ordure. The old tree is, no
doubt, perforated, through and through, by its
winding corridors, as large as your middle finger.
As soon as the vermin had passed this his nonage,
which, as I say, may have occupied a dozen years
at least,* he sank into his short pupa-sleep, and
here we see him paying his first visit to the light
of day.

True; this is his first experience of daylight, and
indeed of anything; for all the pupa-sleep and the
larva-labour were prochronic in this case. The
Beetle is just created.

Hark to that hollow roar! There is no mis-
taking that majestic sound. It is the voice of the
many-sounding sea. Yonder through the trees
we catch a glimpse of its shining face, and here
we are at the verge of the cliffs, against whose

* *B. splendida* has been ascertained to have existed, as an
inmate of the wood of a table, for *more than twenty* years.
(Linn. Trans.; x. 399.)

feet the waves are breaking in white foam. We will clamber down to the rocks.

In this weed-fringed tide-pool there is a fine specimen of the Shore-crab (*Carcinus mœnas*). It is a male just arrived at the perfection of adult age; its carapace smooth and wholly dark-green in hue, its under parts rufous orange. Its claws are large and sharp; and the promptitude with which it presents these formidable weapons, extended to the utmost, shows how conscious it is of its warlike powers.

To all appearance this Crab is several years old;* I mean in this his present perfect or imago form. When this form was first assumed, the diameter of the carapace was not more than an eighth of an inch; it is now two inches; a great many periodical sloughings of the crust must have occurred to accomplish this sixteen-fold increase.

But four distinct metamorphoses were passed before the commencement of this form. There was the Grapsoid form with the outline of the carapace nearly parallel-sided, and the dentations on the sides. Before this there was the Megalopa form,

* The rate of increase in dimensions shown by specimens of this species, now so frequently kept in Aquaria, warrants this assertion; though *how many* years a Crab takes to attain adult size, no exact observations, so far as I know, testify.

with the carapace ovate, and the abdomen projecting behind.   Before this there was the Zoea form,
with the carapace rising into a tall erect spine,
sessile eyes, no claws, and the abdomen a slender
jointed cord ending in a triangular plate.   And
before this, there was the egg, which was laid by
the mother Crab, and carried by her for a considerable time attached to the false feet of her
abdomen.

All these evidences of age, clear and unanswerable though they are, are yet fallacious, because
the Crab has been created but this morning.

On this sea-washed branch of a tree, which has
been blown off by some tempest, and carried into
the ocean, there is a single Barnacle (*Lepas*).   It
consists of a hand of many pairs of fringed fingers,
protected by a shell of five pieces, and a long
flexible cartilaginous stalk, by the lower extremity
of which it adheres to the timber.

The shelly valves are all crossed by strongly
marked lines running over their surfaces in a direction parallel with each other, and with the outer
margins of each valve.   These, like the corresponding foliations in the tube of the *Serpula*, indicate the successive stages of growth; the outlines
of every valve having stood at each of these growth

L

lines in succession.   On each of the scutal valves in this individual I can count about 260 growth-lines : if we suppose one of these to be made in a week,* and the increase to proceed uniformly throughout the year, we must conclude the valve to have been just five years in making.

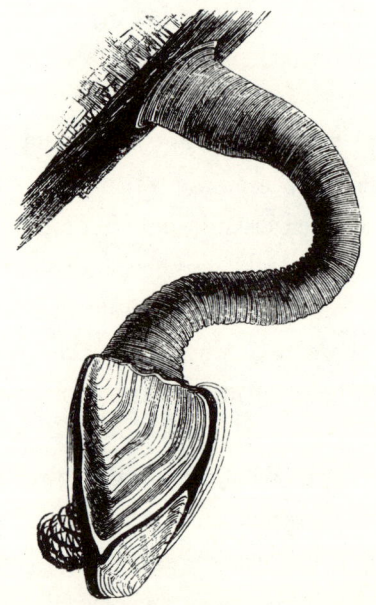

LEPAS.

This animal, like others we have already examined, had, moreover, a history before the first

* The exuvia of the cirri are sloughed from the *Balanidæ* about every week in summer; and perhaps this process is coetaneous with an addition to the valves.

vestige of a valve was formed. It had passed through several metamorphoses; in its pupa stage it had the form of a *Cypris*, and in this condition it first became adherent to the timber: before this it was a larva, having a general resemblance to another Waterflea, the *Cyclops*, especially in its younger stages: in this state it moulted several times. Nor was this the beginning of its life; for there was the still earlier condition common to all these classes of animals, viz. that of the egg, which was laid and carried for some time by the parent Barnacle, and at length hatched while within the valves of her shell.

Thus, through a course of several years we are able to trace back the existence of this Cirriped, to its parent of a former generation. But our conclusions are altogether vitiated by the simple fact that this individual is the first of its species; it never had a parent; it never was an egg.

From the rocky pool before us I have picked up a rough pebble, the surface of which is incrusted with a delicate work of stony lace. This fabric, too fine to be resolved by the unassisted eye, consists of the oval cells of a species of *Lepralia*. There are some hundreds of cells in this patch, which altogether does not cover a square inch of the pebble; and they are all made after one pattern, and set in a very regular manner, in quincunx. Each is a minute slipper-shaped box of stone, with the orifice set round with spines for the protection of the inmate, a transparent, elegant, and sensitive Polypide, which bears on its head a coronet of ciliated tentacles.

I am not going to describe the interesting structure and economy of this atom of life; but merely wish to direct your attention to one point,—the evidence which it affords of the lapse of past time.

Every one of these hundreds of stony cells, together with its living tenant, was normally produced by a process of gemmation; each having budded forth from the side of its predecessor as a knob of clear gelatinous flesh, in the midst of which was developed, first the cell, and then the polypide, —the latter appearing in a rudimentary condition, and gradually acquiring its proper organs, before the orifice of the cell was opened.

I said every one of the cells was thus formed; but I ought to have excepted a single cell, which, though in nowise differing from the rest in form or structure, had a very different origin. This was the primal cell, and its beginning was as follows:

A minute atom of a scarlet hue, and of a semielliptical shape, was one day whirling round and round with rapid gyrations in the open sea. It was of soft consistence, covered with strongly vibrating cilia, and furnished with some stouter setæ. After enjoying its motile instincts awhile, it settled down on this pebble, and became stationary. Presently it secreted and deposited calcareous matter around it, like a coating of the thinnest glass, the red parenchyma receding from the hyaline wall towards the centre.

Soon an orifice with thickened edges appeared on the upper side, and minute spines grew from the edges, which quickly lengthened. It was now a *Lepralia* cell, and now the polypide was developed, and protruded its mouth from the orifice, surrounded by its elegant bell of ciliate tentacles. This solitary cell became the parent of hundreds more, by the gemmative process which I have already described.

But the red swimming atom ;—whence came that ? Well, it was shot out from the interior of a previous *Lepralia*, the result not of a gemmative but of a generative act. It originated in another patch similar to the one which incrusts this pebble, and that, in like manner, and by exactly similar stages, looked back to an anterior patch, and so on.

Plausible as this inference is, it is false; for the little aggregation of cells and polypides has been called into existence by the Divine *fiat*, this very instant.

We are still at the sea-shore. Within the long and narrow crevices into which these low-lying ledges of shale are split, innumerable tufts of sea-weed,—olive, purple, and green,—are perpetually waving in the wash of the sea. On one of these branching shrubs of *Phyllophora,* there is adhering, apparently cast there by accident, an irregular mass of pellucid jelly. It firmly cleaves to the alga, enclosing the bases of several branches within its firm but gelatinous substance.

This knob of jelly is a compound animal of the genus *Botryllus,* and it has just been created as we see it. In order to understand its nature, look at it more closely.

Enclosed in the clear purplish-grey jelly, in the midst of scattered lighter specks, we see several star-like figures of bright hues, in which yellow and red are predominant; the symmetrical arrangement of which pleases the eye, and reminds us of some ornamental pattern designed by human art. Each star is composed of several (three, seven, ten or more) pear-shaped animals, with their smaller ends meeting in the centre around a common orifice, from which a current of water is discharged.

Now this assemblage of animals bears evidence of progressive development. Some time ago a tiny egg was discharged from a parent *Botryllus*, which presently produced a little active tadpole-like larva, called a "spinule." This swam actively by means of its wriggling tail; but at length it settled head downward on this piece of sea-weed. Immediately the head adhered, by an effused cement, to its support; the tail now gradually disappeared; and the round head, in the midst of a mass of jelly-like cement, began to display two orifices on its surface. It soon assumed a pear-like shape, and thus the first *Botryllus* was formed.

From the side of this "pear," another was developed by gemmation, and a third on the opposite side; the smaller ends of all were in contact, and the orifices of these extremities began to merge into one; while the large ends diverged. A fourth and a fifth "pear" were successively produced in the same mode, until a star or "system" was formed. Meanwhile the surrounding mass of living jelly had been commensurately enlarging, and a new *Botryllus*, separate from the other star, had been produced in the jelly, which was the commencing point of a second system; and thus, by degrees, the compound mass of systems has grown to its present state of development.

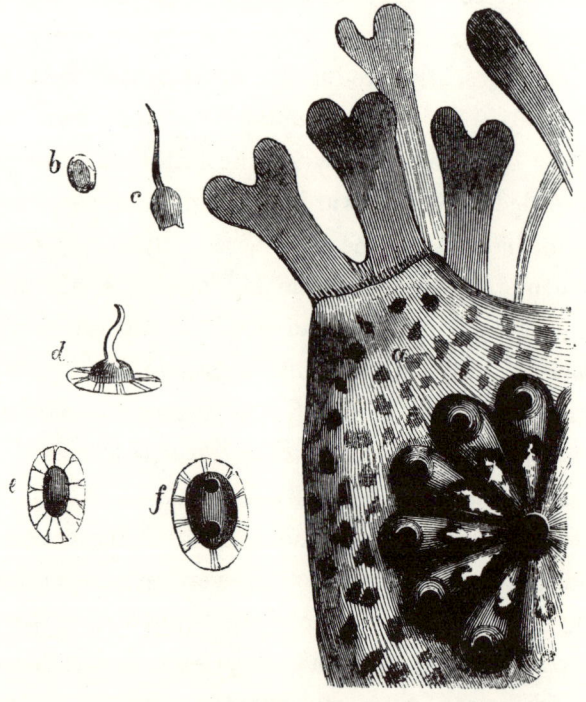

BOTRYLLUS.

a, portion of one system and of a mass, on *Phyllophora rubens;* *b*, an egg *c*, spinule; *d*, the same, attached; *e*, the tail absorbed; *f*, the young *Botryllus.* All magnified.

This process has been one of time : the adhesion of the " spinule " took place in about sixteen hours after its escape from the egg. The appearance of the two orifices was when the little animal was four days old; and by the end of a week a second "pear" had budded. The attainment of the present condition may have occupied about six months.

Nay; time has been no element in this development; it is prochronic development; it is the development of creation, not of nature.

Behold that ruffling of the smooth surface of the water; it is caused evidently by the forcible ejection of a current from some source a little way beneath the surface. Yes, it proceeds from the orifice in this mass of calcareous grit; where the protruding pipe of shell indicates the snug fortress of a *Clavagella.* I will carefully break away a little of the soft stone, and we shall see the curious structure more clearly. Ha! I have split off a piece which nicely exposes the whole burrow, without having materially injured the creature or his shell.

You see it is a bivalve Mollusk with one valve firmly imbedded and cemented into the stony wall of its chamber. But the hinder end of this valve is continued into a shelly tube, intended to protect the siphons, which is carried through the gallery forming the entrance into the chamber, and opens

L 3

by a wide orifice in the free water outside. It is to this tube that I call your attention.

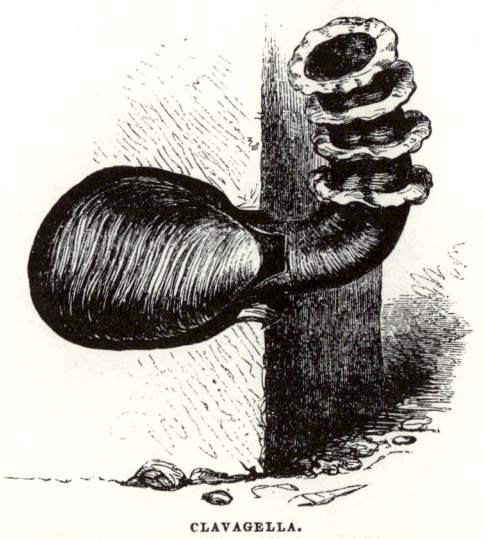

CLAVAGELLA.

You observe that on its outer surface there are several foliated expansions of the shelly substance, surrounding it like so many frills at pretty regular intervals. Each of these foliations is a permanent record of a certain epoch. The terminal one is the margin of the tube-wall everted. The one below this was at some past period the eversion of the margin at what was at that time the extremity. The third frill had in like manner terminated the tube still earlier; and so with the fourth and fifth.

It is impossible to look at these expansions, and not to believe that they have been formed in succession, in this way, by the periodic growth of the tube.

There was a time when the first frill was not commenced; when the creature was a Mollusk with simple valves. But even this was not the beginning of its history. It was as a swimming Infusory with a broad ciliated disk, and a lashing *flagellum*, that the creature commenced its independent career; and it was doubtless in this condition * that it found its way into the burrow of some *Saxicava*. Here its tiny transparent valves were secreted; the left valve was soon cemented to the chamber; and then the creature began to secrete and form the tube around its siphons, which was progressively enlarged, and adorned at every stage of elongation by these witnessing frills—whose testimony is recorded in imperishable stone.

What can be more irresistible than such evi-

---

* Mr. Broderip supposes it to have had the power of swimming freely, and of seeking its future habitation, *as a bivalve;* but Lovèn had not then made known to us the embryogeny and metamorphosis of the *Conchifera*. It is much more probable that the case is as I have ventured to assume in the text.

dence as this? And yet we must take exception to it on the ground that this is the very hour of the animal's creation.

The elegant spinous shell-fish that we discern yonder, half-buried in the sandy floor of the sea— I mean that lilac-tinted Prickly Venus (*Dione Veneris*) needs no shelly protection for its siphons, which, as you may observe, are protruded to a great length. But a lesson not less instructive than that taught by the tube-frills of the *Clavagella* is inculcated by the valves of the *Dione*. Near the hinder margin of each valve there is a ridge which runs from the beak to the front edge, a ridge which

DIONE VENERIS.

bears the series of long slender shelly spines, that imparts such a charm to this shell.

Each of these spines records an interval in the

growth of the shell. There are sixteen distinctly enumerable; each of which may possibly mark a year's growth. The increase of bivalves, however, is slow; and it may be that a longer interval than a year has intervened between spine and spine. For if we look more closely at this beautiful shell, we see that the whole exterior of both valves is marked with concentric foliated ridges, which are also indubitable lines of growth; and that these are twice or thrice as numerous as the spines, from one to five being intercalated between those which support the prolongations of the shelly substance.

Each of these concentric lines has a history. Every line, as well as every spine, has been produced by a protrusion and eversion of the glanduligerous edge of the mantle, which then secreted and poured out a copious deposit of calcareous matter along the margin of the previously existing valve. In this species each periodic deposit took the form of a ridge slightly elevated above the general surface; and, because the turned up margin of the mantle invested the edge of the valve already formed, therefore the new layer, with its elevated ridge, was concentric with the last edge, which was concentric with the previous one, and so on, the common centre of all being the beak (*umbo*) at the back of the valve.

The spines were formed in a manner essentially similar. At every second or third period of increase, the margin of the mantle, which is very versatile and protrusile, was thrust out, at the point which corresponds to the spines, into a long fleshy groove, by the reduplication of its edge. Within this groove the calcareous secretion was poured out; and after it had been allowed a few moments to harden or "*set*," the mantle-groove was cautiously withdrawn, and a new spine was exposed, as a produced end to the foliated ridge.

Yet, though this is the normal and natural mode of production, both of the concentric line and of the spines, it would be illusory to conclude that they have been so produced in the present example. The entire formation of the *Dione* before us has been ab-normal and preter-natural: it has been *created*, not *born*: the whole development so legibly written on the shell has been prochronic.

There goes the Scorpion Stromb (*Pteroceras scorpio*), crawling over the rocks with protruded head and tentacles, and bearing his massive house on his back. This shelly house of his will afford us a good example of structural development.

The great dilated lip, and the long finger-like processes of its edge, had no existence in the youthful days of the shell; they are marks of adult age: when young, the shell was simply spiral, with a thin straight lip bounding a narrow aperture.

Observe also a far more beautiful creature by its side, the Tiger Cowry (*Cyprœa tigris*). Its shell is now entirely enveloped in the meeting wings of the great fleshy mantle, which is mottled with changing hues; and its foot or crawling disk covers a space three or four times as large as the shell. On lifting it in our hand, the whole of this array of soft flesh has been rapidly retracted, and has wholly disappeared within that very narrow orifice, bordered with toothed projections, on the under side of the shell, which we can hardly believe capable of receiving a twentieth part of the bulk that has vanished within it. And now we see nothing but the shell, with its smooth rounded back, marked with dark spots, its white inferior surface cleft by this longitudinal denticulate aperture, and its brilliant porcellanous varnish over the whole.

Now here is evidence of change and progress again. This Cowry-shell is very unlike that of an Olive, with a simple spire, an oval body, a smooth thin lip, and a wide orifice; and as unlike that of a Nautilus. Yet it has passed through both of these stages before it was disguised as we see it now. When it escaped from the egg-shell, it was a minute Pteropod, with two great ciliated disks, inhabiting a transparent nautiloid shell, and swimming giddily about in a revolving fashion. By and by, the tiny shell increased, and the outer whorl lengthened, putting on a long-oval figure. Then—that is, after a considerable period occupied in increasing the dimensions of the shell in this form—it began to assume the adult appearance. The outer lip, which had hitherto been thin, gradually thickened and encroached upon the spire, and the mantle began to secrete and deposit on the outer surface the coat of glassy enamel.

At length the thickening of the lips proceeded to such an extent as almost to conceal the spire, and to reduce the aperture to a narrow line, the edges of which were now thickly plaited with the tooth-like ridges so characteristic of the genus. The lobes of the mantle now protrude through this aperture; and, expanding on each side, have deposited all over the exterior of the shell a coat of glassy enamel, studded with dark round spots or clouds, which entirely conceals the surface with the markings that were formerly visible upon it.

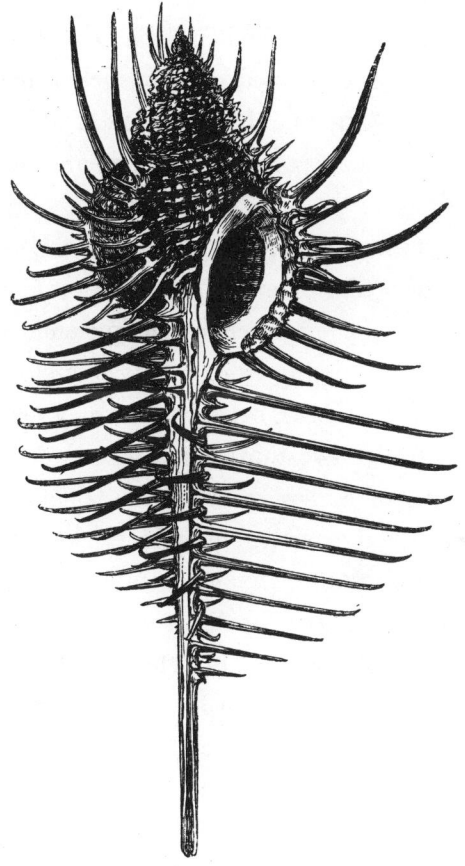

MUREX TENUISPINA.

Yonder Thorny Woodcock (*Murex tenuispina*) is a still more striking shell than either, and one whose periodic growths are peculiarly well marked. It is covered at regular intervals with rows of shelly spines, still longer and more numerous than those we lately admired in the *Dione.* Each series crowns a thickened ridge, which runs across the whorl, as regards the direction of its growth, but longitudinally as regards the general figure of the shell.

Now, the increase of the shell in the Univalves is performed almost exactly as in the Bivalves; namely, by the protrusion and eversion of the mantle on the existing edge. And, therefore, each of these thorny ridges, separated as they are by an interval of just two-thirds of a whorl, marks the termination of a new growth, the shelly matter rising up at the margin in this thickened ridge, which bristles with elongated points.

In this specimen we can trace ten such ridges, whence we legitimately infer ten distinct periods through which this animal has passed, besides the nautiloid stage under which all the creatures of this Class commence existence.

Yet, since each of these three univalves has been this day created, these inferences are deceptive. The Scorpion-shell was *never* otherwise than dilated and digitated. The Cowry has *never* had a lip that was not thickened, nor an exterior that was not porcellánous. The Woodcock has *never* known a moment in which its thorns were less numerous than they are now.

Notice that fine round shell carried along the floor of the sea, by means of a great fleshy tortoise-shell-coloured* body, which, with a head of many spreading tentacles applied to the ground, crawls with a tolerably quick progress.† It is the Pearly Nautilus.

The amplitude of the beautiful nacreous shell is by no means a measure of the dimensions of the animal; for this merely sits within the shallow mouth, like a Welsh fisherman in his coracle. If we remove the creature, we shall find the cavity bounded by a pearly floor, in the centre of which is a slender tube running down from it. On breaking away this floor, we expose an empty chamber, with a similar pearly floor, through which passes the shelly tube, continued through the middle of the chamber, and running down to the next. Thus we should find the whole interior of the shell occupied by a series of these empty chambers, fifty or upwards in number, each less than its predecessor (rather *successor*, if we regard

* Bennett.   † Rumphius.

them in the order of development), until we can trace them no longer in the minute centre of the spire.

Without dwelling on the function of these chambers, farther than to say that they appear admirably contrived to make the animal with its shell either heavier or lighter than the surrounding fluid, by forcing water into them through the tube, and thus condensing the contained air, or by relaxing the pressure, and allowing the elasticity of the air to exclude the water,—our business is just with the formation of the septa, as an evidence of periodic development.*

" The septa are formed periodically, but it must not be supposed that the shell-muscles ever become detached, or that the animal moves the distance of a chamber all at once. It is most likely that the

---

* The periodical formation of these septa in the progress of growth is analogous to that of the projecting external plates in the Wendletrap, and of the rows of spines in the *Murex ;* but those external processes consist of the opake calcareous layer of the shell, whilst the internal processes in the *Nautilus* consist of the nacreous layer, like the septa in the *Turritella*. Thus the embryo *Nautilus* at first inhabits a simple shell, like that of most univalve Mollusca, and manifests, according to the usual law, the general type at the early stage of its existence ; although it soon begins, and apparently before having quitted the *ovum*, to take on the special form.—Prof. Owen's *Lect. on Invertebrate Anim.* p. 593, 2d Ed.

adductors grow only in front, and that a constant waste takes place behind, so that they are always moving onward, except when a new septum is to be formed; the *septa* indicate periodic *rests*." *

These periodic alternations of rest and action, however, it is obvious, can never have really existed in an organism which has but this instant been created. The appearances, therefore, which indicate them, are illusory, considered as testimonies to actual time.

You are aware that what is often spoke of as the "bone" in this Cuttlefish (*Sepia officinalis*), is only a concealed shell; and I need not to dissect the animal to acquaint you that it is a highly interesting structure. A deservedly eminent physiologist shall describe it for us.

"The outer shelly portion of this body consists of horny layers, alternating with calcified layers, in which last may be seen a hexagonal arrangement. The soft, friable substance, that occupies the hollow of this boat-shaped shell, is formed of a number of delicate plates, running across it from one side to the other in parallel directions, but separated by intervals several times wider than the thickness of the plates; and these intervals

* Woodward's "Manual of the Mollusca," p. 83.

are in great part filled up by what appear to be fibres, or slender pillars, passing from one plate or floor to another. A more careful examination shows, however, that instead of a large number of detached pillars, there exists a comparatively small number of very thin sinuous laminæ, which pass from one surface to the other, winding and doubling upon themselves, so that each lamina occupies a considerable space. Their precise arrangement is best seen by examining the parallel plates, after the sinuous laminæ have been detached from them; the lines of junction being distinctly indicated upon these. By this arrangement, each layer is most effectually supported by those with which it is connected above and below; and the sinuosity of the thin intervening laminæ, answering exactly the same purpose as the " corrugation " given to iron plates for the sake of diminishing their flexibility, adds greatly to the strength of this curious texture, which is at the same time lightened by the large amount of space between the parallel plates that intervenes between the sinuosities of the laminæ." *

Now the delicately thin calcareous plates have all been formed in succession, " the first formed

* Carpenter, on the Microscope, &c., p. 602.

being at the outer part and posterior termination of the shell, and the succeeding new layers extending always more forwards than the edges of the old." * They exhibit then many hundreds of distinct deposits, each the result of a separate process; each the work of a definite period of time. The " cuttle-bone " is an autographic record, indubitably genuine, of the Cuttle-fish's history.

Yes, it is certainly genuine; it is as certainly autographic: but it is *not true.* That Cuttle has been this day created.

* Grant's Comp. Anat., 53.

IX.

PARALLELS AND PRECEDENTS.

(*Vertebrate Animals.*)

" The organisation of the body at each epoch may be truly said to be the *resultant* of all the *material* changes which it has undergone during the preceding periods."—*Dr. Carpenter; Human Physiology,* p. 903.

THE *Invertebrata* then agree in one story, and that story is the same as what the plants had told us before. Let us try if the Vertebrate creatures bear them out.

From this promontory we can look far down into the clear profundity of the still and smooth sea. What is that large object that plays hither and thither yonder, now shooting ahead, now resting on his oars, now turning on his course, now cutting the surface, now descending to the depths? It is a full-grown Sword-fish, some ten feet long. We are sufficiently near him to discern that he has one short but high dorsal fin, near the head, and a minute one close to the caudal, the whole intermediate region being smooth. But this is a mark of adult age ; for in early life this same species is furnished with one long and high dorsal, which is continuous from the occiput to the vicinity of the tail-fin. The remotely divided dorsal here tells of many years of life; but tells deceitfully, for the Swordfish is but just created.

Ha! the Swordfish has darted away, like lightning, after a finny victim. See with what doublings and windings he pursues it, and how the terrified prey uses all its powers to escape from its gigantic enemy! Now they near the shore; and now the frightened quarry has leaped out of the sea upon yonder flat shelf of rock, where it lies gasping and floundering, delivered indeed from its pursuer, but only to die by being drowned *in the air*. We will descend from the cliffs, and look at it.

It is a Gilthead (*Chrysophrys aurata*). Life is extinct now; but the brilliant colours and fine metallic reflections are scarcely dimmed—the silvery belly—the azure fins—the sides that gleam like polished steel, inlaid with bands of burnished gold!

I will pluck a scale from this brilliant silvery surface. Its hinder, or free edge, is beset with fine flexible crystalline points, arranged in many successive rows, overlapping each other. The front, or attached edge, is cut in a scolloped pattern, the

M

extremities of undulations that radiate from a common point behind the centre. The whole surface, except the hinder portion that is studded with imbricated points, is covered with an immense multitude of fine concentric lines, which follow the form of the general outline. These are marks of successive increase ; for every one of the lines is the margin of a lamina, the aggregation of which makes up the thickness of the scale. The laminæ can be separated by long maceration in water; and then we see that they are laid one on another in regular order, the uppermost being the smallest, and the first formed ; the last made, which is the largest, being now in contact with the skin.

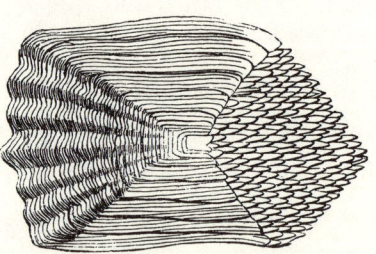

SCALE OF GILTHEAD.

Every scale is therefore a document, on which s indelibly written the record of a multitude of processes, all effected in the past history of the fish. The successively deposited laminæ are exactly analogous to those of calcareous substance in the shell of the bivalve ; * and the evidence is of exactly the same character as what we lately read off from the valve of the *Dione*. But, just as in that example, too, the overruling fact of recent creation precludes our deduction of time from the evidence, since it proves the development to have been prochronic.

I see yonder a more terrific tyrant of the sea than the Swordfish. It is the grisly Shark (*Carcharodon*). How stealthily he glides along, cutting the glittering surface of the sea with his dorsal, and now and then protruding just the tip of the upper lobe of his caudal in the wake of the other! Let us go and look into his mouth; for neither animals nor elements present any impediments to these investigations of ours. Is not this an awful array of knives and lancets? Is not this a case of surgical instruments enough to make you shudder? What would be the amputation of your leg to this row of triangular scalpels, each an inch and a half in diameter? moved, too, by these powerful muscles?

But observe the arrangement of these most formidable teeth. They are not confined to a single row as ours are, but each is succeeded by

* See Jones's General Outline, p. 506. (Ed. 1841.)

another lying behind it, that by another, and
another, and another,—why, there are a dozen
ranks of teeth, lying regularly packed one behind
the other. The object of this arrangement is
a constant supply of new teeth, as those in use
become broken off, or wasted by the sloughing
away of the exterior half-ossified crust of the cartilaginous jaw, to which their base is fastened by
ligaments. Only one row, the outer one, is in use
at once, and this row stands erect; the others lie
flat on each other (more and more completely as
they recede from the outer row); a reserve of
weapons in readiness for use, when those now
employed are done with. There is a continual
growth of the surface to which the teeth are
fastened, from within outwards; so that each of
the reserve rows will in turn be brought to the
edge of the jaw, when it will be thrown up into
the erect position, while the preceding, now turned
out of the mouth by the gradual eversion of the
surface, sloughs away and disappears as an useless
incumbrance. It follows, therefore, that the teeth
which we now see erect and threatening, are the
successors of former ones that have passed away,
and that they were once dormant like those we see
behind them.

But perhaps you may say, What evidence is there that these ever had any predecessors? that they were not originally the front rank as they are now? A very fair question.

In the first place, the great size of the tooth indicates maturity; and is in keeping with the dimensions of the animal,—some twenty feet or so,—which are those of an adult, if not a full-grown individual. But adult age implies previous youth and infancy, and a gradual growth from the length of a few inches to this formidable size. The teeth are found in the embryo Shark when not more than a foot long; and it is evident that many successive generations of teeth have passed away between those pristine lancets of a line in diameter, and these of an inch and a half.

But stay; there is a peculiarity in the structure of these present teeth, which surely indicates their place to be far on in the succession. Each is seen to be finely serrated on its two outer edges,—a provision which, of course, makes them more effective dividers of flesh and bone. But this structure is not found in the teeth of young individuals, which up to a period comparatively advanced, have simply cutting edges.

Hence we are compelled by the phenomena to

infer a long past existence to this animal, which
yet has been called into being within an hour.

On yonder twig sits a beautiful little Tree-frog,
which you would be ready to mistake for a leaf of
more than usually emerald hue, but for the glittering eye, and the line of yellow edged with
purple that passes down the side. Do you notice
the frequent gulpings of the throat? Those are
the periodic inspirations of air, by which the creature breathes; for, having no ribs, by means of
which to depress, and so to expand, the thoracic
cavity, the Frog swallows the air by a voluntary
action. These air-gulps afford us another example
of the sort of evidence we are searching for; they
are so many proofs of a past history. For the
Tree-frog has not *always* swallowed air; there was
a period in its life when it had no lungs; when it
was an aquatic animal, as exclusively a water-
breather as any fish. Fish-like in *form* it was
then, as well as in *habit;* it was a tadpole with
a long compressed muscular tail, and with external
gills of several branches, but as destitute of lungs
as it was of limbs. Any physiologist, looking at
our little green Tree-frog, would pronounce without
hesitation on the stages through which it has
passed; and would describe with the most perfect

confidence the order in which they took place; the gradual absorption of the branchiæ, the development of the lungs, the shrinking up and final disappearance of the tail, the budding forth of the tiny rudimentary limbs, the hinder pair first, then the fore pair, and the subsequent division of their extremities into toes;—the metamorphosis of the little fish into a little batrachian, and the gradual growth and maturation of the latter,—these are facts,—the physiologist would say,—as sure both as to their actuality and as to their order, as that the Frog is a Frog.

Ah! but the physiologist is not aware of a fact, which invalidates all his conclusions based upon experience,—the fact that the little Tree-frog has been created but this very instant.

Hark! that rattling noise is an admonition to us to tread circumspectly. It is the vibration of the horny caudal appendages of a Rattlesnake. And I see the reptile coiled up under yonder shadowing leaf. But our presence is a privileged presence, and so we may handle and examine him with impunity. The organ which produces this sound is composed of a number of hollow horny capsules, each one fitting into the next, in which it is retained loosely by a protuberance of its surface. These, being agitated at the will of the animal, produce that sound which we just now heard. The capsules are developed periodically, one being added to the number already existing every year, until as many as forty are accumulated.* This individual, therefore, having five-and-twenty rattles, must be five-and-twenty years old.

This Snake, however, has had no past years; it has had no yesterday. Its existence commenced this hour.

Here crouches, among the thick reeds, the Leviathan of the rivers, the mailed Crocodile. His body, invested with bony ridged plates, that rise into strong serrations along the tail, seems clothed with power; and his long rows of interlocking teeth, unveiled by lips, appear grinning with perpetual rage. An experienced herpetologist would

---

* Such is the common statement. Dr. Harlan, however, observes that "the rattle is cast annually [with the sloughed skin], and, *consequently*, no inference as to the age of the animal can be drawn from the number of pieces which compose the rattles." (*Journ. Acad. Nat. Sci.*; v. 368.) I confess this appears to me to be a *non sequitur*; for is it not quite possible that one may be added to the *number* annually, without involving the actual perpetuity of the preceding ones? It is evident that the increase must take place at some time or other, and it seems to me more likely to occur at the sloughing of the skin, that is, annually, than either oftener or seldomer.

not fail to find many evidences of age in this huge
reptile. First of all, he would point to its mon-
strous size ; then to the breadth and massive
thickness of the dermal plates. " The head," he
would say, " in the ruggedness of its surface, shows
the same thing, for in youth it was comparatively
smooth ; and also in the form of its outline; for
in this example its length is double its breadth,
whereas in youth, these measurements were nearly
equal. These conical teeth, too, are by no means
the same individual teeth which existed at first.
If you look at the base of one, you will see that it
is hollow, and that the sides of this portion are
already in process of absorption ; that this hollow
cone is a sheath for another tooth beneath, which
is destined to replace it; as this has itself replaced
its predecessor. The large size of the teeth which
we see, therefore, which accords with the dimensions of the jaws, is not a condition induced by
gradual growth, but by a succession of sloughings
and replacements ; and hence the present teeth, in
their size, point conclusively to others which have
preceded them, but which have disappeared."

Yet nothing can be more certain, than that, in
this Crocodile, which has been created to-day, the
successive teeth thus witnessed to, are but ideal,

that is prochronic, teeth; and that all the other indications of the lapse of time, in the development of this individual, are liable to the same exception.

See this solemn, slow-going Tortoise, shut up in his high-domed house of bones. It is the beautiful *Testudo pardalis*, well named from the plates being elegantly spotted and splashed with black on a pale-yellow ground, like the fur of the panther. This is a rather large individual, and the number of concentric lines on the plates of his armour,—or may I not rather say the *tiles* wherewith his house is roofed?—is commensurately great. You see what I mean. Each of the angular plates has a small nuclear lamina, not in the centre of the area, for the development has beén one-sided, but on the highest part. This was the plate in its earliest form, or at least the earliest of which any trace is left; for probably there were others yet earlier and smaller, which, on account of their thinness, have been rubbed away in the travels of the old wanderer. From this nucleus, the plate has been successively enlarged, to correspond with the general growth of the animal, by repeated additions of new laminæ to the inferior surface; each new lamina

being a little wider in every direction than that
which preceded it, though not *equally* on all the
margins ; and thus the plates assumed the form of
a very low cone, as you see, always preserving
the specific outline, and manifesting the stages of
increase, by the projecting edges of the successive
laminæ, exactly as we saw lately in the scales of
the fish.

PLATES OF TORTOISE.

Whether these laminæ are increased in an
annual ratio, I am not sure, nor is it important.
There are, I find, about forty-five concentric lines
on one plate in this specimen, besides others
which are evanescent. Hence it would be quite
legitimate to infer that this Tortoise has passed
through at least forty-five distinct periods of life,

each of which has left a legible record of its existence.

And yet, this moment, in which we look at it, is the very first moment of its life ; the concentric layers are evidences of processes that never occurred, except prochronically.

See yonder stately bird, nearly of the height of man, marching among the luxuriant musa-groves, and feeding on the succulent fruits. There is nothing very admirable in its coarse, black, hair-like plumage; but the rich hues of its naked neck, azure, purple, and scarlet, of the most vivid intensity, attract the gaze. The most remarkable feature in its physiognomy, is the singular, tall ridge of horn on its head, which, like the crested helmet of some mailed warrior, imparts an air of martial prowess to the bird, little in accordance with its peaceful habits.

This protuberance is altogether a development of age. The skull, in the youth of the Cassowary, was scarcely more elevated than that of a chicken; but in the lapse of years, the bony ridge, encased in horn, has gradually elevated itself to the height which it now possesses.

Here again we have a record of time, which is belied by the fact of the bird's recent creation.

What is the glorious train of the Peacock, all filled with eyes, but a false witness of the same kind? It leads us to infer that the bird is three years old at least, since before that period, the covert feathers, which are to form the splendid ornament of maturity, are not developed.

What are the lengthened tail-plumes of most refulgent blue, that adorn the Fork-tailed Humming-bird (*Trochilus forficatus*); what the gorgeously golden tail of the Resplendent Trogon; what the elegant lyre-shaped feathers of the Menura; what the lustrous plumage of the Birds of Paradise,—all of which have been but this hour created,—but so many testimonies, unworthy of confidence, to a past history?

But, further, every individual feather of this beautiful array of plumage concurs in bearing its unblushing witness to the same untruth. What says the physiologist, who is able to read off these autographic records?

" A little while ago, the tips of these feathers were seen each protruding from the extremity of a thick, opaque tube; and a little while before that, the tube itself was a closed capsule, imbedded in a deep follicle of the skin. If you had then cut open the capsule, you would have found two concentric membranous tubes investing a highly vascular secreting pulp, abundantly supplied with nerves and blood-vessels through an orifice at the bottom of the capsule, and destined to form the substance of the coming feather. Indeed, you would have seen the soft, newly-formed barbs folded round the central organized matrix; and below, the incipient quill, filled with the living pulp-cells, and their blood-vessels, which were destined subsequently to wither up and

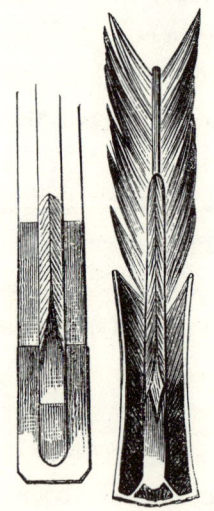

GROWTH OF A FEATHER.

collapse into the light skinny pith which you see in the perfectly matured feather. These are stages which each of these hundreds of feathers has passed through; and these are but a single generation, which have replaced former series that have been lost in the process of moulting, every one of which had in its turn passed through exactly corresponding stages, and so on backward, till we reach the first race of feathers, which were already partly developed when the chick burst forth from its imprisoning egg-shell."

So says the physiologist; but is he not most egregiously in error, since this is the day of these lovely beings' creation?

There goes the great Whale, the true Whale-bone Whale, rolling and wallowing in the trough of the sea, and exposing his enormous black back like an island amidst the white foam, which he stirs up, "making the deep to be hoary." We will use our privilege and take a peep into his mouth, as we did just now into that of the Shark.

What a cavern! and all bristling with long black hair! Why it seems as if the hair grew on the wrong side of his head—on the inside instead of the outside!

Nay, what you call hair is really the Whale's teeth, or what represents teeth. This is the interior free fibrous margin of the *baleen,* which descends in long triangular plates from the upper jaw. There are about two hundred plates on each side, set face to face, with an interval between, and the edges outward. The inward edge runs off into those long hair-like filaments, which also extend from the slender tip. And the whole forms an effective sifting apparatus, by which the volume of sea-water, which the huge creature takes into his mouth in feeding, is drained of the sea-blubbers, the worms, the mollusks, and other small matters, which constitute the subsistence of this vast body.

Now each of these four hundred plates, some twelve· feet in length, has grown from a minute sort of bud, in the upper jaw. Its base is hollow, resting on the formative pulp which is developed from the gum. The pulp is understood to be the immediate origin of the hairy fringe, while a dense vascular substance, seated between the bases of the plates, forms the plate itself. When the plate reaches a certain length, its diameter has become greatly attenuated, and its tip is constantly breaking away, leaving the hair projecting. There is therefore a continual disappearance of the substance of the plates at the tips, and a continual growth at the base to supply the deficiency ; and even more, at least during the period of adolescence, because the actual dimensions of the plates have to be increased in the ratio of the growth of the whole animal.

Here, again, we read a record of past history. The Whale is known to be a long-lived animal ; and a period of many years must have passed in bringing these plates of baleen to their present maturity. Yet the vast organism before us has been created in its vastness but to-day.

On the most prominent shelf of yonder precipice, a sharp buttress of naked limestone, stands an Ibex, guarding, like a watchful sentinel, the herd in the sheltered valley which own his leadership. The pair of noble horns, which are at once his defence and his pride, are marked throughout their ample curve with semi-rings, or knobs, on their anterior side. These afford us an infallible criterion of the animal's age.

We can count in this Ibex fourteen of such prominent bosses. Now the horn in these animals is not shed during life, but consists of a persistent sheath of horny substance, enveloping a bony core. Until full adult age, both the core of bone and the sheath of horn are continually growing; and in the spring, when there is an unusual augmentation of vital energy in the system, the increase is more than usually rapid. At this season, the new matter deposited in the corneous sheath accumulates in the form of one of these bosses, each of which is therefore produced at the interval of a year. As the first boss appears in the second year of the animal's age, we have but to add one to the number of the bosses on each horn, and we have the number of years which it has lived. The Ibex before us is just fifteen years old.

Yon Stag that is rubbing his branchy honours against a tree in the glade,—can we apply the

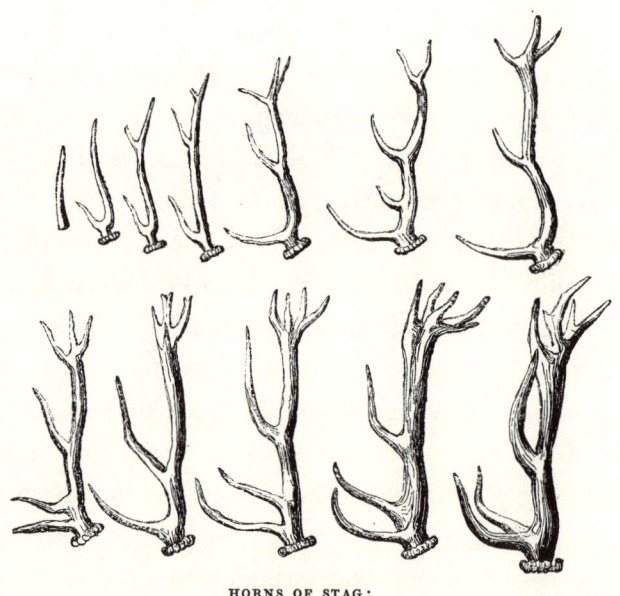

HORNS OF STAG;
In their successive developments.

same criterion to him? Not exactly: for the horns of all the Deer-tribe are of a different structure from those of the *Capradæ*. They are bones of great solidity, not invested with any corneous sheath, but clothed for a certain portion of their duration with a living vascular skin, and are shed every year during life and as constantly renewed.

Yet the bony horns of the Stag are no less sure a criterion of age, at least up to a certain period— than are those of the hollow-horned Ibex. In the spring of the second year of the Fawn, the horns first appear, seated on bony footstalks that spring from the frontal bone. The skin that covers these knobs begins to swell and to become turgid with blood supplied by enlarging arteries. Layers of bone are now deposited, particle by particle, on the footstalks, with surprising rapidity, producing the budding horns, which grow day by day, still covered by the skin, which grows also in a corresponding ratio. This goes on till a simple rod of bone is formed, without any branches. When this is complete, the course of the arteries that supplied the skin is cut off by fresh osseous particles deposited in a thick ring around the base. The enveloping skin then dies, and is soon rubbed off.

After a few months, the connexion of the now dead bone with the living is dissolved by absorption, and the horns fall off.

The next spring they are renewed again, but now with a branch or antler; and the whole falls again in autumn. Every spring sees them renewed, but always with an increase of development; and this increase is definite and well-known; so that the age of a Stag, at least of one in the vigour of life, can be readily and certainly stated.

For example, the individual Stag before us, now browsing so peacefully, has each horn composed of the following elements:—the beam, or main stem; two brow-antlers; one stem-antler, and a coronet of four snags, or royal-antlers, at the summit. This condition is peculiar to the seventh development, to which if we add one year for the hornless stage of fawnhood, we obtain eight years, as, beyond all doubt, the age of this Stag.

Both of these examples, however, the Ibex and the Stag, though so conclusive, and seemingly so irrefragable, are rendered nugatory by the opposing fact of a just recent creation.

See this Horse, a newly created, really wild Horse,

> "Wild as the wild deer, and untaught,
> With spur and bridle undefiled,"—

his sleek coat of a dun mouse-colour, with a black stripe running down his back, and with a full black mane and tail. He has a wild spiteful glance; and his eye, and his lips now and then drawn back displaying his teeth, indicate no very amiable temper. Still, we want to look at those teeth of his. Please to moderate your rancour, generous Dobbin, and let us make an inspection. of their condition!

Now notice these peculiarities. The third pair of permanent incisors have appeared, and have attained the same level as their fellows; all are marked with a central hollow on the crown, the middle pair faintly: the canines have acquired considerable size; they present a regularly-convex surface outwardly, without any marks of grooving on the sides; their inner side is concave; their edges sharp; the third permanent molar has displaced its predecessor of the milk set, and the sixth is developed.*

This condition of the teeth infallibly marks the fifth year of the Horse's age. A year ago the third incisor was only just rising; the canines were small, and strongly grooved, and the third milk grinder was yet existing. A year hence, the central incisors will be worn quite flat, and their marks obliterated; the canines will be fully grown

* Martin "On the Horse," p. 111.

tusks, the second molar will have reached its full
height, and all the teeth will be of the same level.
We can then with perfect confidence assert this to
be a five-year old Horse.   And yet, if we do so,
we shall assert a palpable untruth, for the young
and vigorous stallion has been created to-day.

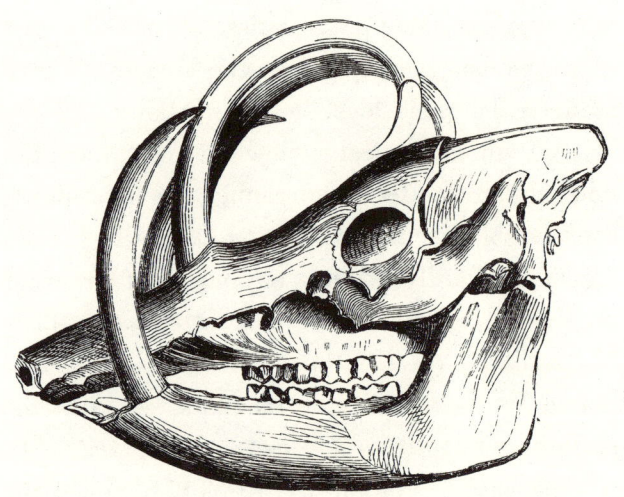

SKULL OF BABIROUSSA.

In the thickets of this nutmeg grove beside us
there is a Babiroussa; let us examine him.   Here
he is, almost submerged in this tepid pool.   Gentle
swine with the circular tusk, please to open your
pretty mouth !   Here are four incisors in the upper
jaw; *at one time there were six.*   The canines of

the same jaw having pierced through the flesh and skin of the face, have grown upward and curved backward like horns; nay, they have nearly completed a circle, and are threatening to re-enter the skull; *once these tusks had not broken from the gums.* There are two premolars : *once there were four.* There are three molars, of which the first is worn quite smooth : *once this surface was crowned with four cones ; but the third molar had not then appeared.*

Away to a broader river. Here wallows and riots the huge Hippopotamus. What can we make of his dentition? A strange array of teeth, indeed, is here; as uncouth and hideous a set as you may hope to see. Yes, but the group is instructive. We will take them in detail.

Look at the lower jaw first. Here are two large projecting incisors in the middle, with their tips worn away obliquely on the outer side, by the action of their opponents in the upper jaw, which are also worn inwardly. The outer incisors, both above and below, are also mutually worn in like manner. The lower canines form massive tusks, curved in the arc of a circle, ground away obliquely by the upper pair; which are short and similarly worn on their front edges. There are three pre-molars

on each side, below and above, much worn: once
there was a fourth, but it was shed early. Lastly,
we find three molars, whose crowns are ground
down so as to expose two polished areas of a four-
lobed figure. A little while ago, these double
areas were trilobate, but at first there were no
smooth areas at all; for these are but sections,
more or less advanced, of the conical knobs, with
which the crown of the molar was originally
armed.*

In both these examples, the polished surfaces of
the teeth, worn away by mutual action, afford
striking evidence of the lapse of time. Some one
may possibly object, however, to this: "What
right have you to assume that these teeth were
worn away at the moment of its creation, admitting
the animal to have been created adult? May they
not have been entire?" I reply, Impossible: the
Hippopotamus's teeth would have been perfectly
useless to him, except in the ground-down condition: nay, the unworn canines would have effectually prevented his jaws from closing, necessitating
the keeping of the mouth wide open until the
attrition was performed; long before which, of

* Professor Owen's "Odontography:"—to which splendid
work I am indebted for the engravings of these skulls.

course, he would have starved. In a natural condition the mutual wearing begins as soon as the surface of the teeth come into contact with each other; that is, as soon as they have acquired a development which constitutes them fit for use. The degree of attrition is merely a question of time. There is no period that can be named, supposing the existence of the perfected teeth at all, in which the evidence of this action would not be visible. How distinct an evidence of past action, and yet, in the case of the created individual, how illusory!

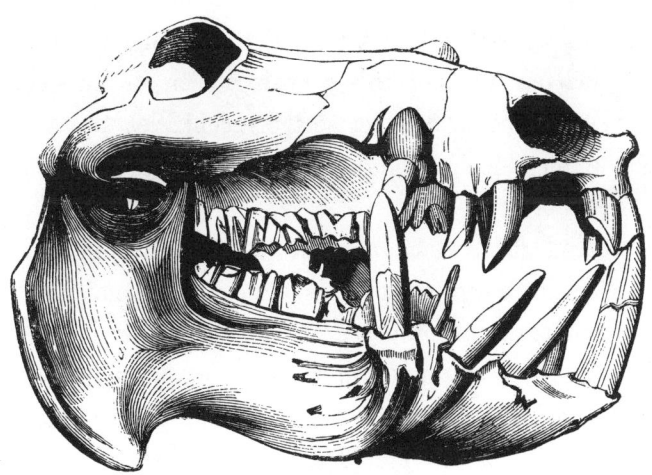

SKULL OF HIPPOPOTAMUS.

N

"Trampling his path through wood and brake,
And canes, which, crackling, fall before his way,
And tassel-grass, whose silvery feathers play
        O'ertopping the young trees,—
    On comes the Elephant, to slake
His thirst at noon, in yon pellucid springs.
Lo ! from his trunk upturn'd, aloft he flings
    The grateful shower : and now
    Plucking the broad-leaf'd bough
Of yonder plane, with waving motion slow,
        Fanning the languid air,
He waves it to and fro."

We will not be content with admiring the vast
size of the fine Dauntelah, and the majesty of his
air and movement, and the intelligence manifested
in all the actions of the "half-reasoning" beast, as
he explores the amœnities of the young world to
which he has but this morning been introduced.
We are out on another sort of scent: let us try if
we can glean any light from him on our present
question.

And, first, we cannot fail to notice his fine pair
of tusks curving upwards almost to a semicircle.
Each tusk is composed of a vast number of thin
cones of ivory, superimposed one on another; ever
increasing by new ones formed within the interior
at the base, and moulded upon the vascular pulp
which fills the cavity, and by which the solid ivory
is constantly secreted and deposited. Each new

cone pushes further and further out those previously deposited, and thus the tusk ever grows in length as it increases in age.

How many years have these tusks occupied in

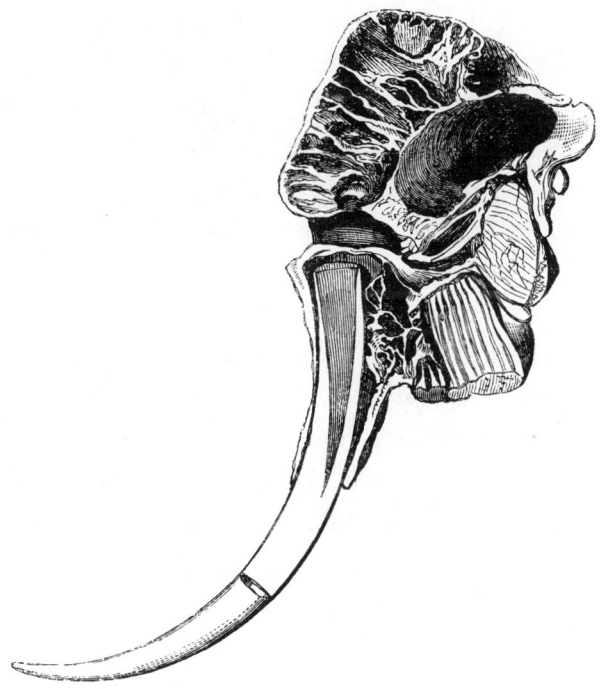

SKULL OF ELEPHANT.

attaining their present diameter and length? We cannot tell: without a transverse section we cannot determine the number of layers of which

N 2

each consists: and if we could, we should yet require to know what ratio exists between the deposition of a cone of ivory and a fixed period of time. The cones, however, in a tusk of these dimensions, are very numerous, for they are but thin; and it is enough for our purpose that they have occupied the same number of periods of time for their formation, though we cannot precisely indicate the length of these periods.

Leaving the tusks, which are the upper incisors, let us now examine the molars. And there is in these a remarkable peculiarity of development, which will assist us greatly in our chronic inquiries. Before we look at them it may be as well to consider this peculiarity.

The Elephant has, from first to last, six, or perhaps eight, molars on each side of each jaw; but there are never more than two partially, or one wholly, in use at once. They have originally an uneven surface, produced by the extremities of a number of what may be considered as so many finger-like constituent teeth, arranged in transverse rows, covered by hard enamel, and cemented together by a bony substance. These points are gradually worn down by the process of mastication, and then the compound tooth appears crossed by

narrow cartouches, or long ovals of enamel, indented at their margins.

" The first set of molars, [*i. e.* the first compound
molar] or milk teeth, begins to cut the jaw eight
or ten days after birth, and the grinders of the
upper jaw appear before those of the lower one.
These milk-grinders are not shed, but are gradually
worn away during the time the second set are
coming forward; and as soon as the body of the
grinder is nearly worn away, the fangs begin to
be absorbed. From the end of the second to the
beginning of the sixth year, the third set come
gradually forward as the jaw lengthens, not only
to fill up this additional space, but also to supply
the place of this second set, which are, during the
same period, gradually worn away, and have their
fangs absorbed. From the beginning of the sixth
to the end of the ninth year, the fourth set of
grinders come forward to supply the gradual waste
of the third set. In this manner to the end of
life, the Elephant obtains a set of new teeth, as
the old ones become unfit for the mastication of
its food.

" The milk-grinders consist each of four teeth,
or *laminæ ;* the second set of grinders of eight or
nine *laminæ ;* the third set of twelve or thirteen ;

the fourth set of fifteen, and so on to the seventh or eighth set, when each grinder consists of twenty-two or twenty-three: and it may be added, that each succeeding grinder takes at least a year more than its predecessor to be completed." *

As each tooth advances, only a small portion pierces the gum at once; one of twelve or fourteen *laminæ,* for instance, shows only two or three of these through the gum, the remainder being as yet imbedded in the jaw; and in fact the *tooth is complete at its fore part,* where it is required for mastication, *while behind it is still very incomplete;* the laminæ are successively perfected as they advance. The molar of an Elephant *can never, therefore, be seen in a perfect state :* for if it is not worn in front, the back part is not fully formed and is without fangs; and when the structure of the hinder portion is perfected, *the front part is already gone.*

· " When the complex molar cuts the gum, the cement is first rubbed off the digital summits; then their enamel cap is worn away, and the central dentine comes into play with a prominent enamel ring; the digital processes are next ground down to their common uniting base, and a transverse

* Brewster's Edinburgh Encyclopædia.

tract of dentine, with its wavy border of enamel, is exposed; finally, the transverse plates themselves are abraded to their common base of dentine, and a smooth and polished tract of that substance is produced. From this basis the roots of the molar are developed, and increase in length, to keep the worn crown on the grinding level, until the reproductive force is exhausted. When the whole extent of a grinder has thus successively come into play, its last part is reduced to a long fang supporting a smooth and polished field of dentine, with sometimes a few remnants of the bottom of the enamel folds at its hinder part. Then, having become useless, it is attacked by the absorbent action, by which, and the pressure of the succeeding tooth, it is finally shed." *

With these physiological facts ascertained, let us proceed to the determination of the actual age of our noble Dauntelah. The molar in present use has a length of about nine inches, and a diameter of three and a half. Its crown is crossed by about eighteen enamel-plates; of which the anterior ones are much worn away, while the hinder ones can scarcely be counted with precision, as they have not wholly cut their way through the gum. These

* Owen's Odontogr. p. 631.

characters indicate the fifth molar (or set of molars) of the whole life-series. And the following facts will help us now to fix the actual age, at least approximately.

The first molar cuts the gum at two weeks old, is in full use at three months, and is shed in the course of the second year. The second cuts the gum at about six months, and is shed in the fifth year. The third appears at two years, is in full use about the fifth year, and finally disappears about the ninth year. In the sixth year the fourth breaks from the gum, and lasts till the animal's twenty-fifth year. The fifth cuts the gum at the twentieth year, is entirely exposed soon after the fortieth, and is thrust out about the sixtieth year, by the advance of the sixth molar, which appears at about fifty years old, and probably lasts for half a century more. If others succeed this,—a seventh and even an eighth, as some assert,—these would carry on the Elephant's life to two or three centuries, in accordance with an ancient opinion, which is in some degree countenanced by modern observations.

To come back, then, to the case before us, since the fifth molar has its fore part much worn, and the posterior laminæ scarcely yet protruded from the gum, it follows that this Elephant is now not far from the fortieth year of his life, a deduction which well agrees with the dimensions of his tusks, and his appearance of mature vigour.

Can you detect a flaw in this reasoning? And yet how baseless the conclusion, which assigns a past existence of forty years to a creature called into existence this very day!

X.

PARALLELS AND PRECEDENTS.

(*Man.*)

" Once, in the flight of ages past,
   There lived a Man,—and who was he?
Mortal, howe'er thy lot be cast,
   That man resembled thee."—MONTGOMERY.

WE have knocked at the doors of the vegetable world, asking our questions; then at those of the lower tribes of the brute creation, and now at those of the higher forms; and we have received but one answer,—varying, indeed, in terms, but essentially the same in meaning,—from all. And now we have one more application to make; we have, still in our ideal peregrination, to seek out the newly-created form of our first progenitor, the primal Head of the Human Race.

And here we behold him; not like the beasts that perish, but—

" Of far nobler shape, erect and tall,
Godlike erect, with native honour clad,
In naked majesty, as lord of all."

The definitive question before us is this : Does the body of the Man just created present us with any evidences of a past existence, and if so, what are they? And that we may rightly judge of the matter, we will, as on former occasions, call in the aid of a skilful and experienced physiologist, to whom we will distinctly put the question.

The Physiologist's Report.

In replying to your inquiry concerning the proofs of a past existence in the Man before me, I must treat of him as a mere animal,—a creature having an organic being.

And, first, I find every part of the surface of his body possessing a nearly uniform temperature, which is higher than that of the surrounding atmosphere. There is, moreover, on all parts of the body, a tinge of redness, more or less vivid in certain regions. The heat, and the carnation tinge, alike indicate the presence of blood, arterial blood, diffused throughout, and, in particular, occupying the capillaries of the superficial parts. Every drop of this blood is preceded and succeeded by other drops, every one of which has been impelled out of the heart by its constant contractions.

But the very existence of this blood supposes the pre-existence of chyle and lymph, out of which it has been constructed. The chyle was formed out of chyme, changed by the action of the pancreatic and biliary secretions. Chyme is food, chemically altered by the action of the gastric juice. So that the blood, now coursing through the arteries and veins, implies the previous process of the reception of food. And these pancreatic and biliary secretions, which are essential to the conversion of chyme into chyle,—and therefore into blood,—do you ask their origin? They were prepared, the one by the pancreas, the other by the liver, from blood already existing,—blood *previously formed* of chyle with the addition of bile, &c.—and so indefinitely.

Again, the blood in these capillary arteries is of a bright scarlet hue, which it derives from its being charged with oxygen. This it received in the *lungs*, parting at the same time with the carbon which it had taken up in its former course. The lungs then must have existed *before* the blood could be where and what it is, viz. arterial blood in the capillaries of the extremities; before it was driven out of the heart, since it was transmitted from the lungs through the pulmonary veins into the heart, thence to be pumped into the arterial system.

But since all the tissues of the body are formed from the blood, the lungs were dependent on already-existing blood for their existence. And as the formative and nutrient power is lodged exclusively in *arterial* blood, the very blood out of which the lungs were organized was dependent on lungs for oxygenation, without which it would have been effete and useless.

Here then is a cycle of which I cannot trace the beginning.

But further. On the extremities of the fingers and of the toes, there are broad horny *nails*. These I trace down to the curved line where they issue from beneath the skin, and whence every particle of each nail has issued in succession. They are composed of several strata of polygonal cells, which have all grown in reduplications of the skin, forming compressed curved sheaths (*follicles*); stratum after stratum of cells having been added to the base-line, as the nail perpetually grew forwards. About three months elapse from the emergence of a given stratum of cells, before that stratum becomes terminal; and therefore each of these twenty-four finger-

and toe-nails is a witness to three months' past existence.

The head is clothed with luxuriant *hair*, composed of a multitude of individual fibres, each of

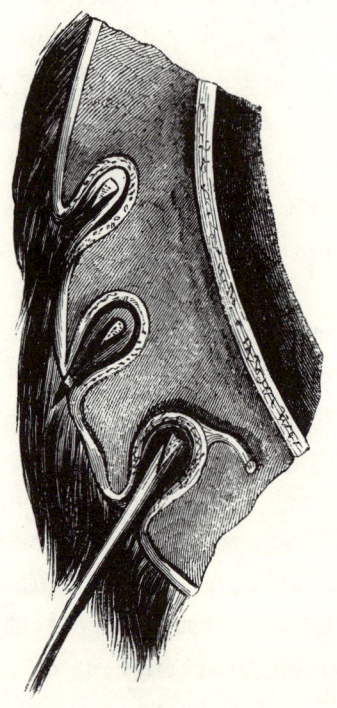

GROWTH OF HAIR (*magnified*).

which is an epidermic appendage, essentially similar to the nails. Every hair is contained at its basal extremity in a delicate follicle, where it terminates around a soft vascular bulb, made up of blood-vessels and nerves. On the surface of this living bulb the horny substance is continually secreted and deposited in layers, each of which in succession pushes forward those previously made, till the tip extrudes from the follicle of the skin, after which it continues to grow in the same way, as an external hair. The tip is gradually worn away; and thus the constant growth cannot, in general, cause it to exceed a certain given length. Each of the thousands of hairs with which this majestic head is clothed, bears witness to past time; and as the increase of hair is about an inch per month, and as this hair is about four inches in length, we have here thousands of witnesses to at least four months of previous history.

The bones which make up the firm and stately fabric about which this human body is built, are no productions of a day. Long before this they existed in the form of cartilages. In these, minute arteries began to deposit particles of phosphate of lime, around certain centres of ossification, doing their work in a determinate order, and in regular lines, so as to form continuous fibres. These fibres, aggregated, and connected by others, soon formed a texture of spicula or thin plates.

Now take as an example a cylindrical hollow bone, as that of the thigh. Here the spicula were arranged longitudinally, parallel to the axis of the bone: preserving the general form of the cartilage which constituted its scaffolding.

But the bone required a progressive increase in size. In its early state, moreover, it was not hollow, but solid. Changes must have taken place to bring it to its present dimensions and condition. These were effected by the actual removal of some parts, simultaneously with the deposition of others.

At a certain stage of ossification, cells were excavated by the action of the absorbent vessels, which carried away portions of bony matter lying in the axis of the cylindrical bone. Their place was supplied by an oily matter, which is the marrow. As the growth proceeded, while new layers were deposited on the outside of the bone, and at the end of the long fibres, the internal layers near the centre were removed by the absorbent vessels, so that the cavity was further enlarged. In this manner the outermost layer of the young bone gradually changed its relative situation, becoming more and more deeply buried by the new layers which were successively deposited, and which covered and surrounded it; until by the removal of all the layers situated near to the centre, it became the innermost layer, and was itself destined in its turn to disappear, leaving the new bone without a single particle which had entered into the composition of the original structure.*

These processes have been the slow and gradual work of years, of the lapse of which years the bones are themselves eloquent witnesses.

Within the mouth there are many *teeth*. I will not now speak of their exact number, nor of some other particulars concerning them, because I mean to return to them presently; but I look only at their general structure and origin. Each tooth consists of three distinct parts, the central portion, which is *ivory ;* the exceedingly hard, polished, glassy coat of the crown, which. is *enamel ;* and a thin layer of bone around the fang, which is the *cement.*

Before either of these appeared, a minute papillary process of vascular pulp was formed in a cavity of the jaw. Over the pulp was spread an excessively thin membrane, which secreted from the blood, and deposited, a thin shell of bony

* Penny Cyclopædia; *art.* BONE.

matter, or ivory, moulded on the form of the
pulp. Successive layers of ivory were then added,
from within; the pulp diminishing in a corresponding ratio. The cavity of the jaw at the same
time deepened, and the pulp lengthened downward

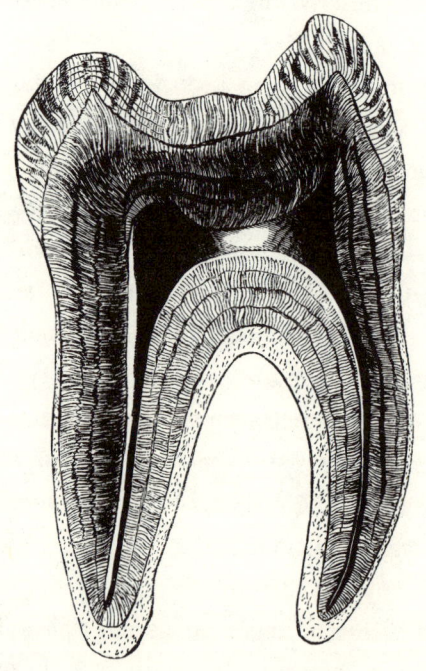

SECTION OF HUMAN TOOTH (*magnified*).

into the space thus provided; layers of bony
substance being gradually deposited upon it, as
above.

The cavity itself was lined with a thick vascular membrane, united to the papilla at its base. Within the space lying between this membrane and the pulp, there was deposited from the wall of the former a soft, granular, non-vascular substance, known as the enamel organ. The cells on the inner surface of this substance then took the form of long, sub-parallel prisms, set in close array, perpendicular to the surface of the tooth. Earthy matter was progressively deposited in them, by which they became the exceedingly dense and hard enamel of the crown. The cement of the fang was then formed by a slight modification of the process which had produced the enamel.

Here, then, are several distinct and important processes, effected in regular and immutable succession, each requiring time for its performance, and all undeniably witnessed-to by the structure of every tooth here seen.

As I have thus proved the *fact* of life existing in this human body for some time previous to the present moment, I now proceed to inquire how far its structure may throw light on the *actual duration* of that past life. How far can we ascertain its chronology?

The stature of the Man before me is about six feet. An infant at birth is from eighteen to twenty-one inches in length. At ten years old the average stature is about four feet. Six feet may be taken as the full adult height of man; and this is attained from the twenty-first to the twenty-fifth year. The stature of this individual would therefore indicate an age not less than twenty-one years.

On the front of the throat I perceive a strongly-marked, angular prominence, formed by the union of the two plates of the thyroid cartilage. The prominence of this angle is due to the enlargement of the larynx; and it is accompanied by a deepening of the pitch of the voice, producing the full rich sounds that we have this instant heard, as the Man chanted his song of praise. These tones, and this projection of the thyroid cartilage, are equally distinctive marks of puberty, and do not appear till about the sixteenth or seventeenth year.

The chin, and sides of the face, are clothed with a dense bush of crisp hair,—the beard. This is a distinctive mark of the adolescent period, and may be taken as indicating an age not less than twenty years.

On again examining the mouth, I find the teeth

are thirty-two in number; viz., four incisors, two
canines, four pre-molars, and six true molars, in
each jaw. None of these existed (at least visibly)
during the first seven years of life; in that period
they were represented by the milk-teeth of infancy. The appearance of the middle pair of
incisors occurred at about the eighth year; the
lateral incisors at nine; the first pre-molars at
ten; the second at eleven; the canines at about
twelve; the second molars at thirteen or fourteen;
and the third molars, or *dentes sapientiæ*, at about
seventeen or eighteen.

The state of the dentition, then, points to an
age certainly not less than the period just named.
How much more it may be, we must gather from
other sources.

I come now to certain phenomena which are
not appreciable to us on mere external examination; but which I am able with certainty to pre-
dicate. And the first of these is the proportion of
arterial to venous blood in the capillaries. In
infancy, the arterial capillaries contain far more
blood than the capillary veins; in old age, the
proportion is exactly reversed; whereas, in maturity, the ratio is just equal. Now, here there is a
very small preponderance of arterial blood, indi-

cating a period but slightly remote from maturity on the side of youth; well agreeing with the conclusion arrived at from previous premises, of some twenty to five-and-twenty years.

Other and more marked manifestations occur in the condition of the skeleton. In the spine, I find *the spinous and transverse processes* of the several vertebræ are completed by separate *epiphyses*, the ossification of which does not commence till after puberty, and the final union of which with the body of the bone does not occur till about the age of twenty-five years.

Each *vertebra*, moreover, has attained a smooth annular *plate* of solid bone, covering a surface that was previously rough and fissured, which is invariably added at the same period.

The *ossification of the sacrum* also has reached its culminating point. At the age of puberty, the component vertebræ began to unite from below upwards, and the two highest have now coalesced; which also marks a period of life not earlier than the twenty-fifth year. The whole united mass, moreover, is furnished on each side with thin bony plates, the appearance of which is no less characteristic of the same age.

Each of the *ribs* is here furnished with two

epiphyses, one for the head and the other for the tubercle ; the ossification of these began soon after puberty ; but their union with the body of the bone, as presented here, has taken several years to accomplish.

To come to the limbs, we find the *shoulder-blade* presenting three *epiphyses,* one for the *coracoid* process, one for the *acromion,* and one for the lower angle of the bone, the ossification of which begins soon after puberty, their union with the body of the bone taking place between the ages of twenty-two and twenty-five years. The *clavicle* has an *epiphysis* at its sternal end, which begins to form between the eighteenth and twentieth years, and is united to the rest of the bone a few years later. The consolidation of the shoulder-bone (*humerus*) is completed rather earlier; the large piece at the upper end, which is formed by the coalescence of the ossific centres of the head and two tuberosities, unites with the shaft at about the twentieth year ; whilst its lower extremity is completed by the junction of the external condyle, and of the two parts of the articulating surface (previously united with each other), at about the seventeenth year, and by that of the internal condyle in the year following. The superior *epiphyses*

of the arm-bones (*radius* and *ulna*) unite with
their respective shafts at about the age of puberty;
the inferior, which are of larger size, at about the
twentieth year. The *epiphyses* of the *metacarpal*
and *phalangeal bones* (those of the hand and
fingers) are united to their principals at about
the twentieth year. In the *Lower Extremities*,
the process of ossification is completed at nearly
the same periods as that of the corresponding
parts of the Upper. The consolidation of the hip-
bones (*ilium, ischium,* and *pubis*) to form the *os
innominatum,* by the ossification of the triradiate
cartilage that intervenes between them in the
socket of the thigh (*acetabulum*), does not take
place until after the period of puberty; and at
this time additional *epiphyses* begin to make their
appearance on the crest of the *ilium,* on its anterior inferior spine, on the tuberosity of the *ischium,*
and on the inner margin of the *pubes,* which are
not finally joined to the bone until about the
twenty-fifth year.*

The concurrence of these conditions in the
skeleton, the nearly balanced ratio of the bloods,
the perfected dentition, the beard, the deepened
voice, the prominent larynx, and the stature, com-

* Dr. Carpenter's Human Physiol. p. 916. (Ed. 1855.)

bine to point out, with infallible precision, the age of this Man, as between twenty-five and thirty years.

So far, then, we can with certainty trace back the history of this being, as an independent organism; but did his history then commence? O no; we can carry him much farther back than this. What means this curious depression in the centre of the abdomen, and the corrugated knob which occupies the cavity?*

This is the NAVEL. The corrugation is the cicatrice left where once was attached the umbilical cord, and whence its remains, having died, sloughed away. This organ introduces us to the

---

* Sir Thomas Browne, indeed, denies Adam a navel; I presume, however, physiologists will rather take my view. Sir Thomas did not know that the prochronism which he thought absurd pervaded every part of organic structure. The following is his verdict:—

"Another Mistake there may be in the Picture of our first Parents, who after the manner of theyre Posteritie are bothe delineated with a Navill: and this is observable not only in ordinarie and stayned peeces, but in the Authenticke Draughts of Vrbin, Angelo, and others. Which, notwythstandynge, cannot be allowed, except wee impute that vnto the first Cause, which we impose not on the second; or what wee deny vnto Nature, wee impute vnto Naturity it selfe; that is, that in the first and moste accomplyshed Peece, the Creator affected Superfluities, or ordayned Parts withoute all Vse or Offyce."—*Pseudodoxia Epidemica*, lib. v.; cap. v.

fœtal life of Man ; for it was the link of connexion between the unborn infant and the parent ; the channel, through whose arteries and veins the oxygenated and the effete blood passed to and from the parental system, when as yet the unused lungs had not received one breath of vital air.

And thus the life of the individual Man before us passes, by a necessary retrogression, back to the life of another individual, from whose substance his own substance was formed by gemmation; one of the component cells of whose structure was the primordial cell, from which have been developed successively all the cells which now make up his mature and perfect organism.

How is it possible to avoid this conclusion? Has not the physiologist irrefragable grounds for it, founded on universal experience? Has not observation abundantly shown, that, wherever the bones, flesh, blood, teeth, nails, hair of man exist, the aggregate body has passed through stages exactly correspondent to those alluded to above, and has originated in the uterus of a mother, its fœtal life being, so to speak, a budding out of hers? Has the combined experience of mankind ever seen a solitary exception to this law? How, then, can we refuse the concession that, in the individual before us, in whom we find all the phenomena that we are accustomed to associate with adult Man, repeated in the most exact verisimilitude, without a single flaw—how, I say, can we hesitate to assert that such was his origin too?

And yet, in order to assert it, we must be prepared to adopt the old Pagan doctrine of the eternity of matter; *ex nihilo nihil fit.* But those with whom I argue are precluded from this, by my first Postulate.

XI.

PARALLELS AND PRECEDENTS.

(*Germs.*)

" Every cell, like every individual Plant or Animal, is the product of a previous organism of the same kind."—(DR. CARPENTER, *Comp. Physiol.* § 347.)

IN the preceding examples I have assumed that every organic entity was created in that stage of its being which constitutes the acme of its peculiar development; when all its faculties are in their highest perfection, and when it is best fitted to reproduce its own image. From the very nature of things I judge that this was the actual fact;*

---

* Blackwood, in an excellent article on Johnston's *Physical Geography* (April, 1849), says :—" Adam *must* have been created in the full possession of manhood; for if he had been formed an infant, he must have perished through mere helplessness. When God looked on this world, and pronounced all to be ' very good,'—which implies the completion of his purpose, and the perfection of his work—is it possible to conceive that he looked only on the germs of production, on plains covered

since, if we suppose the formation of the primitive creatures in an undeveloped or infant condition, a period would require to lapse before the increase of the species could begin ; which time would be wasted. To those, indeed, who receive as authority the testimony of the Holy Scripture, the matter stands on more than probable ground; for its statements, as to the condition of the things created, are clear and full : they were not seeds, and germs, and eggs, and embryos,—but "the tree yielding fruit whose seed was in itself,"—" great whales,"—" winged fowl,"—"the beast of the earth,"—and " man." *

But I do not mean to shield myself behind authority. I have begged the *fact* of creation ; but not the truth, nor even the existence, of any historic document describing it. It is essential to my argument that any such be left entirely out of the

with eggs, or seas filled with spawn, or forests still buried in the capsules of seeds ; on a creation utterly shapeless, lifeless and silent, instead of the myriads of delighted existence, all enjoying the first sense of being ?"

And an eminent Geologist considers the position indisputable, as regards man :—" To the slightest rational consideration it must be evident, that the first human pair were created in the perfection of their bodily organs and mental powers."—(Dr. J. P. Smith ; "Script. and Geol.;" 219.)

* Gen. i. 12, 21, 25, 27.

question ; and, for the present, I accordingly ignore the Bible.

It is possible that some opponent may object to my assumption of maturity in created organisms.

" Your deductions may be sound enough," such an one may say, " provided your newly-created Locust-tree had so many concentric cylinders of timber, your Tree-fern had a well-developed stem of leaf-bases, your Coral a great aggregation of polype-cells, your Tortoise a carapace of many-laminated plates, your Elephant a half-worn set of molars, and your Man a thoroughly ossified skeleton. But how do you know that either of these organisms was created in this mature stage ? I will not deny that each was created,—was called suddenly out of non-entity into entity ; but I believe, or at least I choose to believe,—that each was created in the simplest form in which it can exist ; as the seed, the gemmule, the ovum, the— ahem !"

Pray go on ! you were about to say " the infant," or " the fœtus," or " the embryo," probably ; pray make your selection : which will you say ?

" Well, I hardly know. Because, if I choose the new-born infant, you will say, Its condition implies a nine months' pre-existence, certainly ;

not to speak of the absurdity of a new-born infant being cast out into an open world without a parent to feed it. If I say, The fœtus, or the still more incipient embryo, I involve, at once, a pre-existent mother. I am afraid you have me there!"

I think I have. However, let us take up the matter orderly, and proceed on the supposition that my previous examples must be all cancelled, and the question argued *de novo*, on the assumption that each organism was created in its least developed condition.

It will not be considered necessary, I suppose, to look at any intermediate condition of the organisms. The argument which is based upon the leaf-scales of the Fern or the Palm would essentially apply to either of these plants when it first issues from the ground. At the period when it comprises but a single frond, the botanist would no more hesitate in pronouncing that the organism had passed through stages previous to that one, than he would when it possesses an elongated stipe; though, in the latter case, the evidences of the pre-existence are more patent to the unin-structed eye. He would say, The single frond implies, with absolute necessity, a spore in the one case, a seed in the other; and we need not to see either, to be assured that this must have preceded the leaf-stage.

But you go farther back still. " The plant was created as a seed." Let us renew our imaginary tour at the epoch, or epochs (as many as you please), of creation, on this supposition.

Here is a very young plant of the curious Seychelles Palm or Double Cocoa-nut (*Lodoicea Sechellarum*). A single frond is all that is yet developed, and this is as yet unexpanded, the pinnæ being still folded on the mid-rib, like a fan. Trace the frond down to its base. It springs from a thick horizontal cylindric process, which has also shot down a radicle into the soil. We trace the cylindrical stem along the surface of the soil, and find, lying on the ground, among the grass, but not buried, a great double nut, something like the two hemispheres of a human brain, or like a common cocoa-nut, half split open and healed. Out of this the thick stem has issued; and we find that it is only the cotyledon of the seed, that has prolonged its base in the process of germination, in order to throw up, clear of the nut, the plumule and radicle.

We look at the great nut, and find, on the woody exterior of the fibrous pericarp, at the side opposite to that whence issues the cotyledon, a broad scar. What is this? It is the *mark left by the severance of a footstalk*, which united the fruit to the parent plant. This great drupe was once a small ovary seated in the centre of a three-petaled flower, which, with many others, issued out of a great spathe, a mass of inflorescence, and hung down from the base of the leafy coronal of an adult palm-tree. This scar is an irreproachable witness of the existence of the parent palm.

Here, lying on the dry and dusty earth, is a brown flat bean of great hardness. This is a seed destined by and by to produce that splendid tree *Erythrina crista-galli*. But it has been just created.

This bean bears on one of its edges an oval scar, very distinctly marked, called the *hilum*. This was the point of attachment of a short column, by which the seed was united to one of the sutures of a long pod, in the interior of which it lay, in company with several others like itself. This great legume or pod had been the bottom of the pistil of a papilionaceous flower, crowned by a tiny stigma, lodged in a sheath formed by the united stamens, and surrounded by a corolla of refulgent scarlet petals.

Of course such a flower was not an independent organism; it was one of many that adorned a great tree, the history of whose life would carry us back through several generations of human years.

This single infolding leaf, that is just shooting

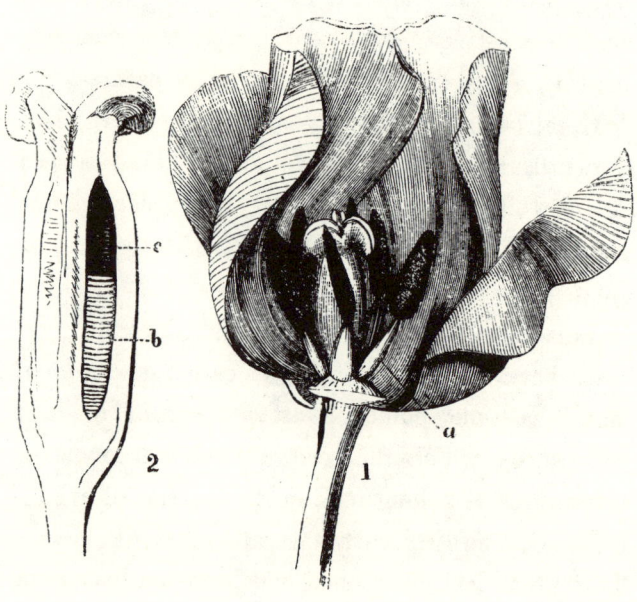

GARDEN TULIP.

Fig. 1. A flower with two petals removed, to show the ovary, *a*.   Fig. 2. The same ovary, more mature, divided longitudinally; *b*, the unripe seeds, packed on each other; *c*, a portion of the same carpel, from which the seeds have been removed.

from the soil, so small and feeble,—what of this? There are certainly no concentric cylinders of timber here : can we trace a previous history of this?

Yes : by carefully removing the soil from the base, we see that it originates in a flat yellow seed —the seed of a Tulip. Here again we have no difficulty in detecting evidence of its former attachment. A great number of these seeds were once closely packed one on another, in each of the three carpels that constituted the capsule. And this capsule had been the oblong, three-sided ovary, which formed the body of the pistil in some beautiful Tulip.

Do you observe these two round fleshy leaves, just peeping from the sandy earth? They are the earliest growths of a plant of *Arachis hypogæa.* In this case again, to understand the true relations of this organism, we must expose it wholly to view.

Beneath the surface of the earth, then, I find that these seed-leaves are the two halves (*cotyledons*) of a kind of pea, which was formerly enclosed in a wrinkled skinny pod. But what is most inter- esting is that the pod is here, the cotyledons shooting out of it. And, attached to one end of the pod, here is a slender stalk, now withered

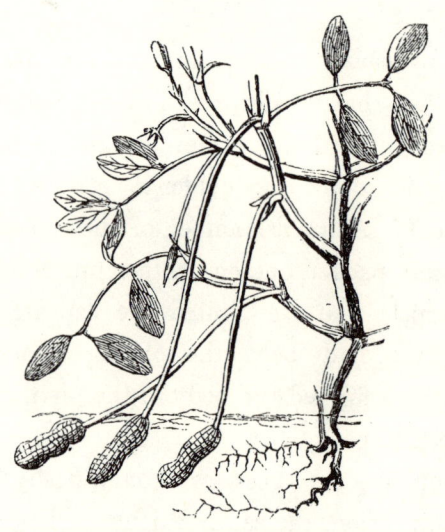

GERMINATION OF EARTH-PEA.

and dry, which projects out of the ground into the air.

Now here we have a beautiful link of connexion with the past. The plant before us does not ripen its seeds, and then drop them to care for themselves, as most plants do. " The young fruit, instead of being placed at the bottom of the calyx, as in other kinds of pulse, is found at the bottom and in the inside of a long slender tube, which looks like a flower-stalk. When the flower has withered, and the young fruit is fertilized, nothing but the bottom of the tube with its contents remains. At this period a small point projects from the summit of the young fruit, and gradually elongates, curving downwards towards the earth. At the same time the stalk of the fruit lengthens, until the small point strikes the earth, into which the now half-grown fruit is speedily forced, and where it finally ripens in what would seem a most unnatural position."*

The young plant before us has been this moment created, and created in this incipient stage of growth: and yet there is, even here, an indubitable evidence, so far as physical phenomena can afford it, of a past history. It would be utterly impossible to select any stage in the life of the Earth-pea, which did not connect itself, visibly and palpably, with a previous stage.

Let us return to the shore-loving Mangrove. You object to my assumption that it was created as a tree, with a well-branched stem elevated

* Penny Cyclop.; *art.* ARACHIS.

upon a series of arching roots; and to my deduction of pre-lapsed years for the formation of those roots. Very well. I give it up. You allow that the primitive Mangrove was created in some stage, but you contend for the germ-stage, the simplest condition of the plant, whatever that might be.

Now, where shall we find it? In the first pair of developed leaves? They certainly point back to the cotyledons. To the cotyledons, then, let us look.

Lo! the young plant is germinating before its connexion with the parent is severed. It is the singular habit of this tree, that its seeds are already in a growing condition, while they hang from the twig. Each seed is a long club-shaped body, with a bulbous base and a slender point, more or less produced. While it yet hangs from the branch, the radicle and crown of the root begin to grow, and gradually lengthen, until the tip reaches the soil, which it penetrates and thus roots itself; while those which depend from the higher branches, after growing for a while, drop, and, sticking in the mud, throw out roots from one end, and leaves from the other.

What have you gained, then, in this case, by going back to the germ? The germ as decisively asserts its origination from an already existing organism — the parent tree—as the flourishing tree witnesses its gradual development from a germ. The Mangrove could not by possibility have been created in any stage, consistent with the identity of the species with that which we behold now in the nineteenth century, — that did not show ocular evidence of a previous history ; — evidence from the nature of things fallacious.

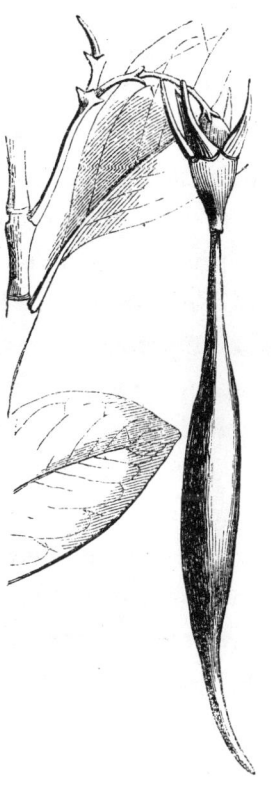

SEED OF MANGROVE.

It would be merely tiresome to go on through the vegetable kingdom. In every plant the simplest condition—viz. that of a spore or seed— depends on some development, or process, or series of processes, that have preceded it. Nor does the lapse of time between the previous process and the

apparent result at all destroy their necessary connexion. In the case of the curious Misseltoes, the
ovule does not appear till three months after the
pollen has been shed; but when it does appear, its
existence as an organism capable of developing
the characteristic form of its species, is as truly
dependent on the previous existence of the pollen,
as if not an hour had intervened.

Supposing the essential conditions of vegetable
organisms to have been at the first what they are
now; in other words, supposing specific identity
to have been always maintained,—which I have
demanded as a postulate for this argument,—it
appears to me demonstrable, that every plant in
the world presented at the moment of its creation
evidences of *prochronic* development, in nowise to
be distinguished from those on which we firmly
rely as proving the lapse of time.

But is the case otherwise in the animal
world?

We traced back the history of our Medusa
through its marvellous series of gemmative developments, till we reached the minute Infusory-like
gemmule, which is its simplest form. Now it is
quite legitimate to assume that *this*, and not the
pulmonigrade umbrelliform stage, was the one in

which the new-created Medusa began existence.
Have we, then, got rid of the evidence of past
time, which we deduced from the successive
changes through which the adult had passed?
What is this ciliated planule, and whence comes
it? It is the embryo discharged from the fringed
ovary of a female Medusa; it has already passed
through several changes of colour and form. It
is now of a deep yellow colour; it has been violet;
it has been colourless: it is now shaped like
a dumb-bell; it was a globule; it had been a
mulberry-mass. Yet earlier, it had been a component cell of the ovarian band, which divided
the generative cavity from that of the stomach, in
the parent Medusa.

In like manner the ciliated gemmule from which
was formed the "pluteus" of the Urchin, was
dependent on the existence of a parent Urchin;
the monadiform germ from which was developed
the pentacrinus of the Feather-star, was originally hidden in the ovarian tubes of a parent
Feather-Star: the infant *Serpula* that deposited
the first atoms of calcareous matter as a commenced tube, had begun its own existence in
the body of a parent *Serpula*.

It is true the evidence of the connexion between

the germ and the parent is not in these low forms always patent to the eye; it is physiological. But it is not less conclusive to one who is able to appreciate its force. A physiologist is as sure that every germ, every ovum, in the Invertebrate animals, was produced by an animal of a former generation, as he is of the same fact in a Mammal, where his eye can see the scar of the umbilical cord.

In many instances there is stronger, or rather more obvious and ordinarily appreciable, evidence of the link between the present and the past generation, than the physiological dependence. The world of Insects, which, from its immensity, and from the high organic rank of its members, affords us so exhaustless a mine of economical wonders,—is rich in examples to the point. A few of these I shall cite.

The eggs of many Insects are not dropped anywhere, at random; for, as the newly-born young have limited powers of locomotion, and yet are in general able to subsist only on some particular kind of food, it is necessary that their birth should occur in the immediate proximity of such food: and therefore that the egg should be so placed. Now this circumstance would not be specially noteworthy if the locality selected for the deposition of the egg were the same as that in which the parent insect had been accustomed to find its own private enjoyments: we should reasonably say that the eggs were placed here, because the parents happened to be here. The case, however, is very different.

We never find the egg of the Peacock Butterfly adhering to the leaf of a cabbage, nor that of the Garden White to the leaf of a nettle; but the nettle is invariably selected for the former, and a cruciferous plant for the latter.

Yet there is nothing in the individual wants or likings of the Butterfly, in either case, to account for this. Both the one and the other flutter through the sunny air, alight to drink the water of some slushy pool, rest on the expanding flowers and probe them for nectar, or suck the exuding juices of an over-ripe fruit. But when did you ever see the gorgeous-eyed Peacock feeding on a nettle, or the White on a cabbage? Eagerly as they seek these plants, it is solely for the purpose of depositing their eggs where instinct teaches them their unborn progeny will find suitable food.

Supposing, therefore, we had found the egg of

either of these butterflies at the moment of its creation, we should assuredly have found it on the nettle or the cabbage (as the case might be); because to suppose it in any other situation would be equivalent to supposing it so placed as that the end of its creation—the life of the species created—would be *ipso facto* frustrated. But, finding it so, the question naturally arises,—Why here, and not elsewhere? and the only possible answer, on the ground of phenomena, is, Because the parent chose this situation for it. And thus we are inevitably thrown back to an anterior generation, which is equivalent to past time.

Again, if we had seen the egg of the Nut Weevil (*Balaninus nucum*) just come from the creative hand of God, we should certainly have found it within the immature soft-shelled hazel-nut, because there alone would the grub when hatched meet with " food convenient for" it. And yet if we had sought (ignorant of the fact of its recent creation) the reason of its being there, our acquaintance with entomology would have pointed to the parent beetle, who, with her jaws placed at the tip of a long slender snout, had bored a tiny hole in the tender shell, and had then projected the egg from her abdomen into the interior.

The eggs of the *Œstridæ*—for example, the Worble of the Ox (*Œstrus bovis*) or the Bot of the Sheep (*Œ. ovis*)—would be discovered in no other circumstances than beneath the skin of the former, and at the edge of the nostrils of the latter. For these are the respective situations in which the egg is always deposited, that of the Worble hatching *in situ*, and forming a superficial abscess in communication with the external air, and that of the Sheep-bot producing a larva which crawls up the nostrils of the poor animal, till it finds a suitable resting-place in the frontal sinuses of the skull. To suppose the egg in any other circumstances than those which I have mentioned, would be to consign it to certain destruction. Yet does not its presence there bear witness to the eclectic care of the parent Gadfly, whose unerring instinct knew how to seek and select the right position?

If you had set yourself to look for the egg of a *Pimpla manifestator*, a common Cuckoo-fly, where would you have looked for it, but in the fatty tissues of a wild bee's grub, that was lodged in a deep hole in some old post? If you had sought elsewhere, you would surely have been disappointed. And would not its presence there bear testimony to the lengthened ovipositor of the well-known brisk and busy fly, and to its remarkable habits?*

The grub of the Pill Chafer or Tumble-dung Beetle (*Phanœus*) feeds on the ordure of *Mammalia*. And, in order that the newly-hatched young may have a copious supply of food at hand, the parent chafer with its jaws detaches a mass of recent ordure, which it then rolls over the ground with its hind feet, until it acquires a globular form, and a coating of earth or sand. An egg is then deposited in the centre of the ball, which is rolled into a hole made in the earth to receive it. The coating of earth drying and hardening, keeps the interior of the mass fresh and moist until the young grub is hatched, when it at once begins to devour its savoury and delicate provision.

It would be vain to search for the egg of a *Cynips* except within a vegetable gall, or at least within the tissues of a plant that are going to produce one. Take as an example *C. quercus*, which produces the spongy excrescence well known as the common Oak-apple. The female Gall-fly is furnished with an ovipositor in the shape of a very fine curved needle, with which she punctures the tender bark of an oak shoot, lodging an

* Linn. Trans. iii. 23.

egg in the perforation. Stimulated by some fluid,
probably, which is poured into the wound at the
same time, the sap forms a peculiar tissue around
the egg, swelling into a large ball, on which the
young grub begins to feed eagerly, and in which
it finds the only nutriment on which it could
subsist.

Now, if we had found the egg of a Gall-fly
newly created, we should certainly have found it
in a gall; and the gall would have afforded us
indubitable evidence of the wounding of the vegetable tissues, and of the organ, secretion, and
instinct of the tiny fly by which the process had
been effected. The evidence would be irresistible,
but of course it would be fallacious.

Let us now look at a few examples in which the
egg is found in invariable association not merely
with something that the parent has found for it,
but with something that has proceeded from her,
a part of herself.

Of this nature are the eggs of that beautiful, but
most cacodious, lace-winged fly, *Chrysopa perla*.
If you had seen one of these (or more) at the
instant of its creation, you would have seen a tiny
oval body placed at the extremity of an elastic
footstalk half-an-inch in length; and as fine as

a hair, standing erect from the surface of a leaf.
This thread is composed of a gummy secretion,

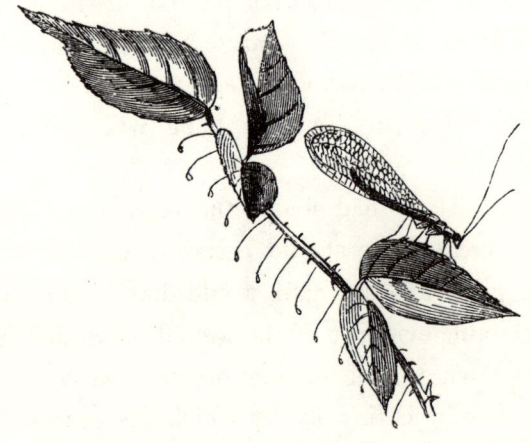

LACE-FLY AND EGGS.

evolved in a gland attached to the oviduct of the
female Lacefly. When she deposits an egg, she
first exudes a drop of this gum on the surface of a
leaf, and then, elevating her abdomen, the viscid
substance is drawn out in a thread, which presently hardening in the air, the egg is left at the
tip of the filament. An experienced entomologist,
on seeing this object, would have no hesitation in
declaring the origin of the footstalk to be the gum-
gland of the female *Chrysopa;* and yet he would
certainly have drawn a false inference in the case
that I am supposing.

Many Spiders enclose their eggs in an envelope, the produce of their own bowels. Take an interesting example, as narrated by the eloquent Mr. Kirby. "There is a Spider common under clods of earth (*Lycosa saccata*), which may at once be distinguished by a white globular silken bag, about the size of a pea, in which she has deposited her eggs, attached to the extremity of her body. Never miser clung to his treasure with more tenacious solicitude than this spider to her bag. Though apparently a considerable incumbrance, she carries it with her everywhere. If you deprive her of it, she makes the most strenuous efforts for its recovery; and no personal danger can force her to quit the precious load. Are her efforts ineffectual? a stupefying melancholy seems to seize her; and, when deprived of this first object of her cares, existence itself appears to have lost its charms. If she succeeds in regaining her bag, or you restore it to her, her actions demonstrate the excess of her joy. She eagerly seizes it, and with the utmost agility runs off with it to a place of security.

"The attachment of this affectionate mother is not confined to her eggs. After the young spiders are hatched, they make their way out of the bag

P

by an orifice which she is careful to open for them, and without which they could never escape; and then, like the young of the Surinam toad (*Rana pipa*), they attach themselves in clusters upon her back, belly, head, and even legs; and in this situation, where they present a very singular appearance, she carries them about with her, and feeds them until their first moult, when they are big enough to provide their own subsistence."*

I waive the argument derived from the fact of the apparent necessity of the mother's care for the new-born young. But the mother's care is indispensable to the appearance of the young at all; not only because the eggs are the produce of her ovary, but also because the envelope which protects them is the produce of her spinning-glands.

There is a furry moth, by no means uncommon, known to collectors as the Gipsy (*Hypogymna dispar*), the eggs of which require to be protected by an elaborate covering, either from extremes of temperature, from light, or from certain electric conditions of the atmosphere. The protection is afforded at the expense of the hair which clothed the mother herself. Her ovipositor is furnished

* Introd. to Entom. ; Lett. xi. § 2.

with a pair of nippers, by means of which she plucks off her own hairs, and makes with them a flat cushion on the surface of a leaf. On this she deposits her eggs in successive layers; and when the full number is laid, she covers them with a roof of hair, slanting downwards and outwards from an apex, so artfully arranged, like the thatch of a cottage, as effectually to throw off water; each layer of hairs overlapping the preceding, and all preserving the same direction, so that, when finished, the work resembles a smooth and well brushed piece of fur.

If, then, a patch of eggs newly-created had been subjected to our inspection, we should have found them snugly protected by their conical roof of thatch; and when we came to examine the thatch microscopically, we should have found it composed of the hairs of *Hypogymna*. And thus again we should have an indubitable and yet deceptive record of a preceding existence.

The numerous species of the genus *Coccus*, to which we are indebted for cochineal, lac, and other products valuable in commerce, afford me an illustration of my argument, more striking than any of the above. In the case of the lac insect (*C. lacca*), for example, the female resembles a little hemispherical scale on the twig of a tree. At a certain period of her life, a pellucid, glutinous substance begins to exude from the margins of her body, which by and by completely covers it, cementing her firmly to the branch, from which she never afterwards moves. She now proceeds to lay her eggs, which one by one as they are extruded are thrust under her, between her abdomen and the surface of the branch. The result of this is, that when the whole are laid, they occupy pretty nearly the same position in relation to the mother as they did before, with this exception, that the abdominal integuments, which before were beneath them, are now above them, and are in close contact with those of the back, so-that both together make a double, but still a thin, arched roof over the heap of eggs, which are thus protected till the hatching of the young, when they eat their way out of their long dead mother.

Let me now make my usual application. You say the *Coccus* was created not an adult insect, which would involve the prochronic stages of its metamorphosis, but as a germ, that is an egg (for the germ of an insect is an egg, and nothing else) : well, here is a batch of Coccus-eggs just created, covered with the scaly roof which is necessary to

their existence. But this scale is not a record of
the mother, but the mother herself, *a prochronic
mother,* of course!

Other genera of this wonderful class of animals
yield us evidences of a somewhat different character, in the structures which the parents form for
the reception of their eggs.

One of the most complex and elaborate pieces
of mechanism found in any animal organ is the
ovipositor of the Sawflies (*Tenthredinidæ*). I cannot here describe it at length; it may suffice to
say that it consists of two saw-plates, working
separately and in opposite directions, the teeth of
which are cut into finer teeth; and two supporting
plates, very similar to the saws in shape and
appearance. The whole flat side of the saw is,
moreover, covered with minute sharp points, which
give the action of a rasp to the instrument, in
addition to that of saw.

By means of this complicated apparatus the
parent fly cuts a groove in the twig of the proper
shrub, say, a rose-bush. When it is made, the
plates are slightly separated, and an egg is laid in
the groove. The saw is now withdrawn, and a
frothy secretion is deposited, which appears to be
intended, by its hardening, to prevent the growth

of the wood from closing upon the egg, before the time of hatching arrives.

If, then, any of the species of *Tenthredo* had been called into primal existence as an egg, it must have been within such a groove as this; and the groove, if carefully examined, would have presented evidences of having been formed and filled by the curious implement of the parent fly.

Those obscure and obscene Insects, the Cockroach tribe (*Blattadœ*), secrete an extraordinary covering for the protection of their eggs. "Instead of being laid separately, the eggs are, when deposited, enclosed in a horny case, or capsule, variable in its form in different species, but generally of a more or less compressed oval shape, resembling a small bean. There is a longitudinal slit in the margin of the capsule, each side of which is defended by a narrow serrated plate, fitting closely to its fellow. The inside of this egg-case is divided into two spaces, in each of which is a row of separate compartments, every one enclosing an egg, so that the whole resembles the pod of some leguminous vegetable. This capsule, with its precious contents, is constantly carried about by the female for a week or a fortnight, and is then fastened, by means of a glutinous fluid, in some safe locality.

The slit of the capsule is strongly coated with cement, so as to be even stronger than the other parts. In this capsule the young larvæ are hatched, and immediately discharge a fluid which softens the cement, and enables them to push open the slit, through which they escape; after their exit the slit shuts again so closely, that it appears as entire as before. In some species it would seem that the females themselves liberate their offspring by seizing the capsule when the larvæ are fit for escape, and tearing it with the aid of their fore-legs from end to end, by which means the enclosed larvæ are set at liberty." *

It is impossible to read this description without being reminded of the manner in which the bean or other leguminous seed links itself with a former generation by means of the dehiscent legume, itself a production of the parent plant. And the same reasoning applies to this case, as to the other;—the egg, if the *Blatta* was created in that stage, would triumphantly show in the pod with which it was covered, a record of past processes.

So, once more, with the immense tribes of solitary Bees, Wasps, and Spheges. I shall mention

* Jones; Nat. Hist. Anim.; ii. 151.

but one example, from my own experience. It is the Dirt Dauber (*Pelopœus flavipes*) of North America. The female of this elegant fly, when about to lay her eggs, builds up a tubular nest of cells with fine mud, which she makes by mingling and kneading road-dust with her saliva. Each tube consists of several cells, separated by transverse partitions of the mortar; and in each, before she closes it up, she lays a single egg, which she then covers with spiders which are to constitute the food of the grub when hatched, and to last it during the whole period of its larval growth. Dead spiders would not do, for their bodies would either dry up, or become putrescent long before the young grub could devour them. On the other hand, if a number of these fierce and carnivorous creatures were immured, in health, they would soon destroy one another. To obviate this, the parent-fly ingeniously stings every spider just sufficiently to paralyse, without killing it. Thus nearly a score of living spiders are packed away in a cell scarcely larger than a lady's thimble; and thus they remain fresh and succulent food for the larva, not only till it is ready to begin its eating task, but even to the close of its repast.

I think this a particularly instructive example.

The *Pelopœus* was indubitably créated; for it exists. As indubitably it was created in some stage of its cyclical life-history. If as an imago, then I press the argument from the necessity of its previous metamorphoses. If as a pupa, or a larva, or an egg, each of these conditions of life was entirely passed as an inmate of the mud-walled cottage; which cottage was built and stocked with food by the industry and skill of the parent-fly. The grub could not have lived without the stored spiders; the spiders could not have been stored (*normally*) without the agency of the fly.

In some other instances the connexion between germ and parent is patent to the eye. The beautiful Starfish, *Cribella*, passes through all its infant metamorphoses, changing from an ovum to an Infusory, thence to a Pluteus (or what is analogous to it), thence to a Starfish, all in the marsupium provided for the occasion, by the drawing together of the arms of the patient mother. The female *Brachionus* carries its deposited eggs attached to the hinder part of its body; and thus we can trace, through their transparent coats, the gradual development of the organs of the embryo,—the coloured eye, the rotatory cilia, the complex mastax,—and even detect the vigorous movements of these and other parts, while yet carried hither and thither by the parent.

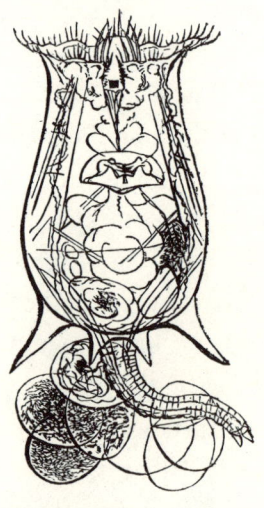

FEMALE BRACHIONUS, WITH EGGS.

But further, in the class from which I have taken this last illustration—that of the ROTIFERA—there are examples of viviparous genera; and these, because of the perfect transparency of all the integuments, are peculiarly instructive and germane to my argument.

In *Rotifer macrurus* the ovary with its germinal vesicles is distinctly seen occupying one side of the animal. From this one of the vesicles enlarges, until it becomes a long-oval translucent sac, nearly

filling the whole left side of the visceral cavity.
A kind of spasmodic movement is suddenly observed in this oblong ovum, and instantly we see,
in its place, a well-developed living young; as
distinctly visible as if it were excluded. It lies
in a bent position, with its foot upturned; is nearly
half the length of the parent; is furnished with a
proboscis, with a pair of crimson eyes, with ciliary
wheels, with a mastax whose toothed hemispheres

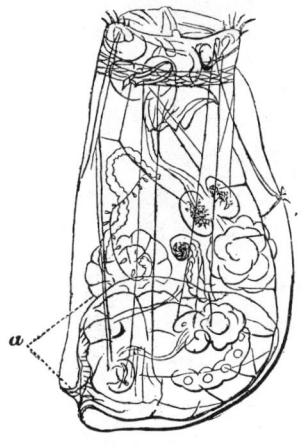

PREGNANT ASPLANCHNA.
a. Unborn young.

frequently work vigorously, and with all the
viscera proper to the species.

In the beautiful, comparatively large, and econo-
mically singular genus, *Asplanchna*, the same

process of development can be watched with perfect facility through every stage.

In the body of the female parent, as transparent as the clearest glass, the band-like ovary is seen floating in the visceral cavity, with several ova in various degrees of advancement. We trace one of these till it becomes a manifestly living young in the ovisac, lying along at the bottom of the parental cavity, more than one-third of whose volume is occupied by it:—supposing it to be a female infant. All its organs,—the eyes, the jaws, the stomach, the pancreatic glands, *the ovary with its nuclei*, the muscles, the rotatory cilia, &c. can be traced with the utmost distinctness long before birth, and its motions are strong and voluntary.

Neither in this case, nor in that of *Rotifer*, does the young animal pass through any metamorphosis; the unborn young has the full development of the parent, in every respect but size. In each case, the *visible* life-history of the individual commences not at birth, but at a period long antecedent, if indeed it can be said to commence at all, where we see it gradually developed from a nucleus, which was an integral part of the parental ovary, *even before that parent's birth.*

In the case of the amusing little Water-fleas (*Daphnia*), we have another example of viviparous generation, which, owing to the same cause as in the ROTIFERA,—the transparency of the integument, can be followed through all its stages by the eye of the observer. The eggs of this little Crustacean are deposited in a special chamber within the valves of the parent, where they are hatched. The young remain in their receptacle for a period, which varies according to the temperature, but long enough for them to undergo important changes in structure, and to pass their first moult.*

Here, again, it is impossible to select a condition which does not take hold of a pre-existence; for the youngest independent stage is dependent on earlier stages; and these are passed in visible connexion with the parent.

It is true there is in this genus, another mode of reproduction, by means of eggs which are thrown off enveloped in an organic covering, called the ephippium. If this condition be selected for the argument of my supposed opponent, I reply that it amounts to nearly the same thing; only the case will then come into the category of those animals

* Cf. Mr. Lubbock (Proc. Roy. Soc. viii. 354), with Dr. Baird (Brit. Entomostr. p. 82).

whose earliest stages are protected by coverings formed from the body of the parent,—like the *Hypogymna*, and the Cockroach, already alluded to.

Where then, in these species, can we possibly select a stage of life, which is not inseparably and even visibly connected with a previous stage?

If we come to the vertebrate creatures, the argument becomes assuredly not less convincing. The formidable Shark, which we considered as a well-toothed adult ready for slaughter, let us suppose to have been created in the harmlessness of infancy. It is a slender thing, some ten or twelve inches long, bent upon itself, inclosing in the ring thus made, the vitellus or yelk-bag, the contents of which are in process of being absorbed into the abdomen. But the whole,—Shark, yelk-bag, and all—is imprisoned in a brown horny capsule, that looks like a pillow-case, with long tapes appended to the four corners.

This very peculiar protecting capsule points clearly to a peculiar structure in the parent. The embryo was not inclosed in the pillow-case, at its first formation; but, in the course of its descent from the ovary through the oviduct, it had to pass a region of the latter, where was a thick glandular mass,—the nidamental gland,—whose office it was to secrete a dense layer of albumen, with which the embryo became invested. This substance took the form of the flattened purse, or pillow-case, with produced angles, above described, and on its exclusion from the duct assumed a very tough horny consistence, and a dark mahogany colour.

The comparative anatomist would, therefore, without the least hesitation, refer the origin of the investing capsule to the nidamental glands of the female Shark; but supposing the embryo to be but just created, his physiological science would only lead him to a false conclusion.

If the Tree-frog afforded us evidence of pre-existent time, in the metamorphosis which it must naturally have experienced from the tadpole to the reptilian condition, what shall we say to that strange and uncouth member of the same class,— the Surinam Toad (*Pipa*)? Little would be gained by selecting the germ-stage, as the presumed epoch of creation in this case; for, according to the extraordinary economy of this genus, the male acts as midwife, and the female as wet-nurse, to the hopeful progeny.

" As fast as the female deposits her eggs, the male who attends her arranges them on her broad back, to the number of fifty or upwards. The contact of these eggs with the skin appears to produce a sort of inflammation.; the skin of the back swells, and becomes covered with pits or cells, which enclose each a single egg, the surface of the back resembling the closed cells of a honey-comb. The female now betakes herself to the water; and in these cells the eggs are not only hatched, but the tadpoles undergo their metamorphosis, emerging in a perfect condition, though very small, after a lapse of *eighty-two days* from the time in which the eggs were placed in their respective pits."

To a tyro in animal physiology it might seem that the smooth rounded egg of a bird or a lizard, presents an example of an organism in the simplest possible condition, and in a stage which, if any can be, is independent of anything that went before.

But is it so? Let us see. Here is the egg of the common Fowl. I take it in my hand, and perceive nothing but an uniform, smooth, hard, white surface. This I break, and find that it is a thin layer of calcareous substance, which, on microscopical examination, proves to be composed of minute polygonal particles, so agglutinated as to leave open spaces in the interstices of their contiguous angles.

Below this calcareous shell I find a membrane (*membrana putaminis*), which seems, from its thinness in most parts, to be single, but which is separated into two layers at the large end of the egg.

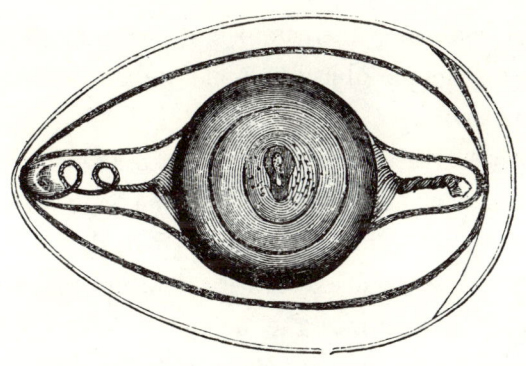

HEN'S EGG.

Within this membrane there is another (*the chalaza*) which, closely enveloping the yelk, passes off from it towards each extremity of the egg in the form of a twisted cord.

Then comes a delicate membrane (*memb. vitelli*) in close contact with, and enveloping the orange-coloured yelk; which latter carries, on one point of its globular surface, the thin *blastoderm* or germinal membrane.

The yelk-globe, fastened by its twisted *chalazæ*, is suspended in a glairy fluid (albumen), which fills the space between it and the *membrana putaminis*. This fluid, though apparently homogeneous, is really composed of many layers, and the innermost of these it is which is condensed into the *chalaza*.

Such, then, is the complex structure of this apparently simple object. What light can it throw on our inquiry?

Each of these component parts bears witness to a succession of past periods. The yelk with its germ was first formed, escaping naked, or clothed only with its own excessively delicate membrane, from its ovisac into the oviduct. Through the course of this tube it now slowly descended, receiving successive investments as it proceeded. The albumen was deposited layer upon layer from the mucous membrane of the upper part of the oviduct; the first depositions condensing into the *chalaza*. By and by it came down to a region of the oviduct where a tenacious secretion was poured out, which, investing the albumen, soon hardened into a substance resembling thin parchment, and formed the *membrana putaminis;* two successive layers of this were deposited, between which a bubble of gas, chiefly composed of oxygen generated in the interval, was inclosed. Then it descended still farther, to a part where the lining membrane of the duct was endowed with the power of secreting calcareous matter, which, as above stated, was deposited in a thin layer of polygonal atoms. And now, having received all its components, and having arrived at the orifice of the duct, the egg was laid.

Here, then, there is abundant evidence of successive processes, which must have preceded the existence of this complete and perfect egg. But there is yet one more evidence which I have reserved to the last, because it is peculiarly distinct and palpable, even to the senses.

The *chalaza*, we see, is twisted at each pole of the yelk-globe, until it resembles a piece of twine : what is the meaning of this ? It was, as I observed, deposited as a loosely enveloping membrane in the upper part of the oviduct; the yelk-globe, however, was progressively descending; and, as it descended, *it continually revolved upon its axis ;* by means of which rotation the investing membrane was gathered at each pole into a spirally twisted cord, stretching from the yelk to the ends of the *membrana putaminis.* Thus it presents us with an unmistakeable record of what took place in the earlier periods of the descent.

We saw distinct traces of the past in the structure of a feather. But the feathers have already begun to develop before the young bird leaves the egg. And the structure of the egg carries us back to the oviduct of the parent-fowl.

At what stage of existence, then, could a bird, by possibility, have been created, which did not present distinct records of prochronic development?

If we come to the MAMMALIA, the impossibility of finding such a stage becomes only more and more obvious. For it is a law in physiology, that the higher the grade of organization assigned to any being, the more it is assisted in infancy by the parent.

" This law is remarkably exemplified in the class MAMMALIA, which unquestionably ranks at the head of the animal kingdom, in respect to degree of intelligence and general elevation of structure. It is the universal and most prominent character-istic of this class, that the young are retained within the body of the female parent, until they have made considerable progress in their development; that, whilst there, they derive their support almost immediately from her blood; and that they are afterwards nourished for some time by a secretion which she affords." *

The fœtus of the Kangaroo, when expelled from the womb, is scarcely more than an inch in length. Its limbs and its tail are indeed formed, but the imperfect creature has been compared to an earthworm, for the colour and semi-transparency of the integument. In this condition it is unable to find and seize the nipple, and equally unable to draw sustenance therefrom, by its own unaided efforts. The *milk is ejected*, by the *muscular action of the mother*, into the throat of the fœtus, and there is a peculiar and beautiful contrivance to obviate the danger of the injected fluid's passing into the trachea instead of the œsophagus.

Yet, from this helpless naked condition to that of the active, well-clothed, experienced young, able to quit the maternal pouch at will, and flee to it for protection, there is a well-understood and perfectly appreciable concatenation of stages, each of which looks back to, and depends on, those previously existing. And, during the whole of these, the mother's presence is necessary to the comfort,

* Dr. Carpenter : Comp. Phys.; p. 615.

and, for the greater part of them, to the very existence of the infant.

Thus, once more, there is no condition of the animal, on which we may fix; as being so simple, as to have no retrospective history.

The umbilical cicatrix I have already alluded to; but I may be permitted to mention it again; because, in all the higher MAMMALIA, at least, it exists, throughout life, an eloquent witness to the organic connexion of the individual with a mother, and therefore to her pre-existence. If it were legitimate to suppose that the first individual of the species Man was created in the condition answering to that of a new-born infant, there would still be the need of maternal milk for its sustenance, and maternal care for its protection, for a considerable period; while, if we carry on the suggested stage to the period when this provision is no longer indispensable, the development of hair, nails, bones, &c., will have proceeded through many stages. And, in either condition, the navel cord or its cicatrix remains, to testify to something anterior to both.

XII.

THE CONCLUSION.

"We have no experience in the creation of worlds."
                                                CHALMERS.

WE have passed in review before us the whole organic world: and the result is uniform; that no example can be selected from the vast vegetable kingdom, none from the vast animal kingdom, which did not at the instant of its creation present indubitable evidences of a previous history. This is not put forth as a *hypothesis,* but as a *necessity ;* I do not say that it was *probably* so, but that it was *certainly* so; not that it *may have been thus,* but that it *could not have been otherwise.*

I do not touch the inorganic world: my acquaintance with chemistry is inadequate for this: perhaps the same law does not extend to the inorganic elements: perhaps their developments and combinations are not, like the economy of plants and animals, essentially and exclusively

cyclical : perhaps carbon and oxygen and hydro-
gen could be created in conditions, which obviously
did not depend on any previously existing conditions.   This I do not know : I neither affirm nor
deny it.   But I think I have demonstrated in these
pages, that such a cyclical character does attach to,
and is inseparable from, the history of all organic
essences ; and that creation can be nothing else
than a series of irruptions into circles : that, supposing the irruption to have been made at what
part of the circle we please, and varying this condition indefinitely at will,—we cannot avoid the
conclusion that each organism was from the first
marked with the records of a previous being.   But
since creation and previous history are inconsistent
with each other ; as the very idea of the creation
of an organism excludes the idea of pre-existence
of that organism, or of any part of it ; it follows,
that such records are *false*, so far as they testify to
time ; that the developments and processes thus
recorded have been produced without time, or are
what I have called *prochronic*.

Nor is this conclusion in the least degree affected
by the actual chronology of creation.   The phenomena were equally eloquent, and equally false,
whether any individual organism were created six

thousand years ago, or innumerable ages; whether primitively, or after the successive creations and annihilations of former organisms.

The law of creation supersedes the law of nature; so far, at least, as the organic world is concerned. The law of nature, established by universal experience, is, that its phenomena depend upon certain natural antecedents : the law of creation is, that the same phenomena depend upon *no* antecedents. The philosopher who should infer the antecedents from the phenomena alone, without having considered the law of creation, would be liable to form totally false conclusions. In order to' be secure from error, he must first assure himself that creation is eliminated from the category of facts which he is investigating; and this he could do only when the facts come within the sphere of personal observation, or of historic testimony. Up to such a period of antiquity as is covered by credible history, and within such a field of observation as history may be considered fairly cognisant of,—the inference of physical antecedents from physical phenomena, in the animal or vegetable world, is legitimate and true. But, beyond that period, I cannot safely deduce the same conclusion; because I cannot tell but that at any given

Q

moment included in my inquiry, creation may have occurred, and have been the absolute beginning of the circular series.

The question of the actual age of any species, whether plant or animal, is one which cannot be answered, except on historic testimony. The sequence of cause and effect is not adequate to answer it; for a legitimate use of this principle, supposing it the only element of the inquiry, would inevitably lead us to the eternity of all existing organic life.

One of the familiar street-exhibitions in the metropolis is a tiny coach and horses of glittering metal; which, by means of simple machinery, course round and round the margin of a circular table. Let us suppose two youths of philosophical turn to come up during the process. They gaze for a while, and one asks his companion the following question.

"How long do you suppose that coach has been running round?"

"How long! for an indefinite period, for aught I know. I have counted twenty-two turns, and can see no change: nor can I suggest any point where the course could have begun."

Here a shrewd lad, carrying a grocer's basket, breaks in.

" Oh no ; there have been only six-and-twenty turns altogether. Four turns had been made when you came up. The whole began by the man taking the carriage out of a box; then he set it down out there, just opposite to us, and gave it a little push with his finger, and it has been running ever since. I saw him do it."

Now perhaps you will say that a glance at the machinery beneath the table would show in a moment how many turns had been made, and how many could be made. Very true: but what if the tramp had locked up his clock-work, and would not let you look at it?

The only evidence worth a rush is that of the lad who saw the whirligig set a-going.

I wish it to be distinctly understood that I am not proving the exact or approximate antiquity of the globe we inhabit. I am not attempting to show that it has existed for no more than six thousand years. I wish this to be distinctly stated, because I am sure I shall meet with many opponents unfair enough, or illogical enough, to misrepresent or misunderstand my argument, and sound the trumpet of victory, because I cannot demonstrate *that*. *All* I set myself to do, is to invalidate the testimony of the witness relied on for the indefinitely remote antiquity; to show that in a very large and important field of nature, evidence exactly analogous to that relied on would inevitably lead to a false conclusion, and must, therefore, be rejected, or received only contingently; received only as indicative of probability, and that only in the absence of any positive witness to the contrary.

Perhaps it may be objected, that there is no sufficient analogy between the phenomena from which the past history of a single organism is inferred, and those from which the past history of a world is inferred. Is there not?

Permit me to repeat an illustration I have already used. The geologist finds a fossil skeleton. His acquaintance with anatomy enables him to pronounce that the objects found are bones. He sees cylinders, condyles, cavities for the marrow, scars of attachment of muscles and tendons, foramina for the passage of nerves and blood-vessels; he finds the internal structure, no less than the form and surface, such as to leave not a doubt that these are real *bones*. Now universal experience has taught him that bones imply the existence of flesh; that flesh implies blood; that blood implies life; that life implies time. He therefore concludes unhesitatingly, that this skeleton was once alive,

and that time passed over it in that living condition.

Is not this process of reasoning exactly parallel
to that which he would have pursued if he had
examined an animal the moment after its creation,
(supposing this fact to be unknown to him,) and by
which he would in like manner have inferred past
time? And where is the vital difference between
the two cases, which would operate to make a conclusion which is manifestly false in the one case
necessarily true in the other?

One of the most eminent of living botanists has
set forth in striking terms the parallelism which I
am suggesting. Speaking of the *shoot* as the vegetable individual, and the woody trunk as a kind
of ever-accumulating ground, which supports successive generations of shoots, he uses the following
comparison.

" The history of the grand development of nature
on the surface of our globe presents an analogy,
which may perhaps serve to set this relation in a
clear light. The successive geological formations
superposed during the course of countless ages,
present, buried in their depths, the traces of as
many formations of the organic world, each of

which carpeted the then superior stratum of the earth with a new life, until it found its own grave in the succeeding formation, when a new uprising of organic life took its place. In the same way, the stem of a tree is a multistratified ground, in whose layers the history of earlier growths is legibly preserved. The number of the woody layers indicates the number of the generations which have perished, *i. e.* the age of the whole tree: a distinct annual ring is the monument of a vigorous season, an indistinct one of a bad season, a sickly one (which is often found among healthy ones) indicates the unhealthiness of the foliage of that particular year. The practised woodman can decipher many facts of the past in the layers of the trunk; *e. g.* a good season for foliage or for seed, damage by frost or by insects, &c."*

In order to perfect the analogy between an organism and the world, so as to show that the law which prevails in the one obtains also in the other, it would be necessary to prove that the development of the physical history of the world is circular, like that already shown to characterise

* Dr. Alex. Braun, "On the Veget. Individual." (Ann. N. H. Nov. 1855.)

the course of organic nature. And this I cannot prove. But neither, as I think, can the contrary be proved.

The life of *the individual* consists of a series of processes which are cyclical. In the tree this is shown by the successive growths and deaths of series of leaves: in the animal by the development and exuviation of nails, hair, epidermis, &c.

The life of *the species* consists of a series of processes which are cyclical. This has been sufficiently illustrated in the preceding pages, in the successive developments and deaths of generations of individuals.

We have reason to believe that species die out, and are replaced by other species, like the individuals which belong to the species, and the organs which belong to the individual. But is the life of *the species* a circle returning into itself? In other words, if we could take a sufficiently large view of the whole plan of nature, should we discern that the existence of species δ necessarily involved the pre-existence of species γ, and must inevitably be followed by species ϵ? Should we be able to trace the same sort of relation between the tiger of Bengal and the fossil tiger of the Yorkshire caves, between *Elephas Indicus* and

Elephas primigenius, as subsists between the leaves of 1857 and the leaves of 1856; or between the oak now flourishing in Sherwood Forest and that of Robin Hood's day, from whose acorn it sprang? *

I dare not say, we should; though I think it highly probable. But I think you will not dare to say, we should *not*.†

It is certain that, when the Omnipotent God proposed to create a given organism, the course of that

---

* It may be objected that *Elephas primigenius* is absolutely distinct from *E. Indicus*. I answer, Yes, *specifically* distinct; and so am I distinct from my father,—*individually* distinct. But as individual distinctness does not preclude the individual from being the exponent of a circular revolution in the life-history of the species, so specific distinctness may not preclude the species from being the exponent of a circular revolution in some higher, unnamed, life-history.

† " We may assert of the individual, as well as of the species, that it completes the cycle of its existence in a succession of subordinate generations; while, on the other hand, we may affirm of the species, that, like the individual, it exhibits a determinate cycle of development." "The species itself may be regarded as an inferior 'momentum' of a still more comprehensive cycle of development."—*Dr. A. Braun*, " *On the Vegetable Individual.*"

" The species is an individual of a higher rank."—*Link : Elements of Botanical Science*, vi. 11.

" Species, like individuals, have a certain limited term of existence. It is the fact, that, *according to some general law*, species of animals are introduced, last for a limited period, and are then succeeded by others performing the same office."— *Ansted's Ancient World*, 52, 54.

organism was present to his idea, as an ever revolving circle, without beginning and without end.
He created it at some point in the circle, and gave
it thus an arbitrary beginning; but one which involved all previous rotations of the circle, though
only as ideal, or, in other phrase, prochronic. Is
it not possible—I do not ask for more—that, in
like manner, the natural course of the world was
projected in his idea as a perfect whole, and that
He determined to create it at some point of that
course, which act, however, should involve previous stages, though only ideal or prochronic?

All naturalists have speculated upon the great
plan of Nature; a grand array of organic essences,
in which every species should be related in like
ratio to its fellow species, by certain affinities, without gaps and without redundancies; the whole constituting a beautiful and perfect unity, a harmonious
scheme, worthy of the infinite Mind that conceived
it. Such a perfect plan has never been presented
by any existing fauna or flora; nor is it made up
by uniting the fossil faunas and floras to the recent ones; *yet the discovery of the fossil world has
made a very signal approach to the filling up of the
great outline;* and the more minutely this has been
investigated, the more have hiatuses been bridged

Q 3

over, which before yawned between species and species, and links of connexion have been supplied which before were lacking.*

It is not necessary,—at least it does not seem so to me,—that all the members of this mighty commonwealth should have an actual, a diachronic existence; any more than that, in the creation of a man, his fœtal, infantile, and adolescent stages should have an actual, diachronic existence, though these are essential to his normal life-history. Nor would their diachronism be more certainly inferrible from the physical traces of them, in the one case than in the other. In the newly-created Man, the proofs of successive processes requiring time, in the skin, hairs, nails, bones, &c. could in no respect be distinguished from the like proofs in a Man of to-day;

---

* "The unity of the plan of organization, and the regular succession of animal forms, point out a *beginning* of this great kingdom on the surface of our globe, although the earliest stages of its development may now be effaced : and the continuity of the series though all geological epochs, and the *gradual transitions* which *connect* the species of one formation with those of the next in succession, distinctly indicate that they form *the parts of one creation,* and not the heterogeneous remnants of successive kingdoms begun and destroyed: so that, while they present the best records of the changes which the surface of the globe has undergone, they likewise afford the best testimony of the recent origin of the present crust of our planet, and of all its organic inhabitants."—*Dr. Grant, in Br. Sci. Annual for* 1839.

yet the developments to which they respectively testify are widely different from each other, so far as regards the element of time. Who will say that the suggestion, *that the strata of the surface of the earth, with their fossil floras and faunas, may possibly belong to a prochronic development of the mighty plan of the life-history of this world,*—who will dare to say that such a suggestion is a self-evident absurdity? If we had no example of such a procedure, we might be justified in dealing cavalierly with the hypothesis; but it has been shown that, without a solitary exception, the whole of the vast vegetable and animal kingdoms were created,—mark! I do not say *may* have been, but MUST have been created—on this principle of a prochronic development, with distinctly traceable records. It was *the law of organic creation.*

It may be objected, that, to assume the world to have been created with fossil skeletons in its crust,—skeletons of animals that never really existed,—is to charge the Creator with forming objects whose sole purpose was to deceive us. The reply is obvious. Were the concentric timber-rings of a created tree formed merely to deceive? Were the growth lines of a created shell intended to deceive? Was the navel of the created Man intended to deceive him into the persuasion that he had had a parent?*

These peculiarities of structure were inseparable

---

* Dr. Harris has the following observations :—

"Why might not God have created the crust of the earth, just as it is, with all its numberless stratifications, and diversified formations, complete? And the analogy for such an exercise of creative power is supposed to be found in the creation of Adam, not as an infant, but as *an adult;* and in the production of the *full-sized* trees of Eden. To which the reply is direct : the maturity of the first man, and of the objects around him, could not deceive him by implying that they had slowly grown to that state. His first knowledge was the knowledge of the contrary. He lived, partly, in order to proclaim the fact of his creation. And, could his own body, or any of the objects created at the same time, have been subjected to a physiological examination, they would, no doubt, have been found to indicate their miraculous production in their very destitution of all the traces of an early growth ; whereas the shell of the, earth is a crowded storehouse of evidence of its gradual formation. So that the question, expressed in other language, amounts to this : Might not the God of infinite truth have enclosed in the earth, at its creation, evidence of its having existed ages before its actual production? Of course, the objector would disavow such a sentiment. But such appears to be the real import of the objection; and, as such, it involves its own refutation."—*Pre-Adamite Earth*, p. 83.

Now this reasoning appeared, doubtless, very triumphant to the worthy Doctor : and yet a very little acquaintance with physiology would have taught him that he was enunciating an absurdity. The very supposition which he considers as self-refuting, is an indubitable physiological fact. I have abundantly shown, in the text, that the *cells which compose* the tree or the animal are as undeniable evidences of past processes as the concentric cylinders of timber, or the superposed layers of bone and scale, from the adult stage of these creatures respectively, without which they would not have been what they were. The Locust-tree could not have been an adult *Hymenœa*, without concentric rings;— nay, it could not have been an exogenous tree at all. The *Dione* could not have been a *Dione* without those foliations and spines that form its generic character. The Man would not have been a Man without a navel.

To the physiologist this is obvious; but some unscientific reader may say, Could not God have created plants and animals without these retrospective marks? I distinctly reply, No! not so as to preserve their specific identity with those with which we are familiar. A Tree-fern without scars on the trunk! A Palm without leaf-bases! A Bean without a hilum! A Tortoise without laminæ on its plates! A Carp without concentric lines on its scales! A Bird without feathers! A Mammal without hairs, or claws, or teeth, or bones, or blood! A Fœtus without a placenta! I have indeed written the preceding pages in vain, if I have not demonstrated, in a multitude of examples, the absolute necessity of retrospective phenomena in newly-created organisms. But if it can be undeniably shown in one single example, our failure to perceive it in ninety-nine other instances would in nowise invalidate the deduction from that one. Granted that you can triumphantly convict me of a *non-sequitur*, in ninety-nine out of every hundred of the cases in which I have attempted to show this connexion; still, if I have conclusively proved that in one solitary instance an animal or a plant was created with but one solitary evidence of pre-development, the principle for which I contend is established.

I trust, however, it does not rest on one example, nor on twenty, nor on a hundred. It may be thought that I have multiplied my illustrations needlessly: ten times as many might have been given. I wished to show that the proof is of a cumulative character: a single good example would, indeed, have established the principle; but I wished to show how widely applicable it is; that it is, indeed, of universal application in the organic kingdoms.

If, then, the existence of retrospective marks, visible and tangible proofs of processes which were prochronic, was so necessary to organic essences, that they could not have been created without them,—is it absurd to suggest the *possibility* (I do no more) that the world itself was created under the influence of the same law, with visible tangible proofs of developments and processes, which yet were only prochronic?

Admit for a moment, as a hypothesis, that the Creator had before his mind a projection of the whole life-history of the globe, commencing with any point which the geologist may imagine to have been a fit commencing point, and ending with some unimaginable acme in the indefinitely distant future. He determines to call this idea into actual existence, not at the supposed commencing point, but at some stage or other of its course.* It is clear, then, that at the selected stage it appears, exactly as it would have appeared at that moment of its history, if all the preceding eras of its history had been real. Just as the new-created Man was, at the first moment of his existence, a man of twenty, or five-and-twenty, or

---

* I here assume the life-history of the globe to be represented by a straight line, because I cannot *prove* it to be a circle. I cannot even *imagine* its circularity. I do not mean the possibility;—I can imagine *that :* but the *mode* I cannot conceive. This, however, does not disprove the possibility. If man's science extended not beyond the accumulated observations of his own life, he would probably be quite incompetent to conceive how the life-history of such a tree as the Oak could be a circle; if he had never seen more than one individual, which was a tree when he was born, and continued to flourish till his death.

thirty years old; physically, palpably, visibly, so old, though not really, not diachronically. He appeared precisely what he would have appeared had he lived so many years.

Let us suppose that this present year 1857 had been the particular epoch in the projected life-history of the world, which the Creator selected as the era of its actual beginning. At his fiat it appears; but in what condition? Its actual condition at this moment:—whatever is now existent would appear, precisely as it does appear. There would be cities filled with swarms of men; there would be houses half-built; castles fallen into ruins; pictures on artists' easels just sketched in; wardrobes filled with half-worn garments; ships sailing over the sea; marks of birds' footsteps on the mud; skeletons whitening the desert sands; human bodies in every stage of decay in the burial-grounds. These and millions of other traces of the past would be found, *because they are found in the world now;* they belong to the present age of the world; and if it had pleased God to call into existence this globe at *this* epoch of its life-history, the whole of which lay like a map before his infinite mind, it would certainly have presented all these phenomena; not to puzzle the philosopher, but because they are inseparable from the condition of the world at the selected moment of irruption into its history; because they constitute its condition; they make it what it is.

Hence the minuteness and undeniableness of the proofs of life which geologists rely on so confidently, and present with such justifiable triumph, do not in the least militate against my principle. The marks of Hyænas' teeth on the bones of Kirkdale cave; the infant skeletons associated with adult skeletons of the same species; the abundance of coprolites; the foot-tracks of Birds and Reptiles; the glacier-scratches on rocks; and hundreds of other beautiful and most irresistible evidences of pre-existence, I do not wish to undervalue, nor to explain away. On the hypothesis that the actual commencing point of the world's history was subsequent to the occurrence of such things in the perfect ideal whole, these phenomena would appear precisely as if the facts themselves had been diachronic, instread of prochronic, as was really the case.*

* The existence of Coprolites—the fossilized excrement of animals --has been considered a more than ordinarily triumphant proof of real pre-existence. Would it not be closely parallel with the presence of fæces in the intestines of an animal at the moment of creation? Yet this appears to me demonstrable. It

Perhaps some one will say, " All this might be tenable, supposing the world were an organism. Your argument goes to show that organic essences in every stage of their existence present proofs of pre-existence ; but what analogy is there between the lifeless inorganic globe (in which evidences of past processes are apparent, independent of the fossil organisms), and a living organic being,— plant or animal?"

I answer, The point in the economy of the organic creatures, on which their prochronism rests, is not the organic, but the circular condition of their being. The problem, then, to be solved, before we can certainly determine the question of analogy between the globe and the organism, is this:—Is the life-history of the globe a cycle? If it is (and there are many reasons why this is pro-

---

may seem at first sight ridiculous, and will probably be represented so ; but truth is truth. I have already proved that blood must have been in the arteries and veins of the newly-created Man (*vide* p. 276, *supra*), and that blood presupposes chyle and chyme ; but what became of the indigestible residuum of the chyme, when the chyle was separated from it ? Would it not, as a matter of course, be found in the intestines ? If the principle is true, that the created organism was exactly what it would have been had it reached that condition by the ordinary course of nature, then fæcal residua must have been in the intestines, as certainly as chyle in the lacteals, or blood in the capillaries.

bable), then I am sure prochronism must have been evident at its creation, since there is no point in a circle which does not imply previous points. At all events, geologists cannot prove that it is not.

Wherever we can discern a cyclical condition, there the law of which I am treating must hold good; and it certainly obtains in other things beside organisms. When the inorganic crust of the globe was first cleft to contain rivers, whence came the water that flowed through the fissures? A river is the produce of rivulets, which issue from mountain springs; these originate in the water that percolates through the soil; and this is derived from the rains, and snows, and dews, that are deposited from the atmosphere. But there would be no deposition from the atmosphere if the water had not first been carried up by evaporation; and the vaporable fluid is obtained from the moistened soil; from the lakes and rivers; and from the seas and oceans, whose loss is perpetually recruited from the flowing rivers. Here, then, we get a circle closely analogous to that of organic being. Was a given drop of water created as a component particle of a running stream? Its position and condition looked back to the mountain spring whence it must naturally have issued. Was it called into being in the spring? It looked up to the surface, whence it must have oozed. Was it formed on the surface? It looked to the clouds, whence it must have dropped. Was it created in the cloud? It looked down to the surface of the lake or sea, whence it must have been raised. Was it created in the lake? It looked to the river, whence it must have flowed.

The chief pelagic currents, which have hitherto so often been the destruction of the navigator, but which may yet become his able and subject servants, flow in circular systems. There is such an one in the southern part of the Indian Ocean, known as the Hurricane Region; another immense one ever running round and round the North Pacific; and, above all, that wondrous river of hot water—a river whose well-marked banks are not solid earth, but cold water—the Gulf Stream.

" The fruit of trees belonging to the torrid zone of America is annually cast ashore on the western coasts of Ireland and Norway. Pennant observes that the seeds of plants which grow in Jamaica, Cuba, and the adjacent countries, are collected on the shores of the Hebrides. Thither also barrels of French wine, the remains of vessels wrecked in the West Indian seas, have been carried. In 1809

His Majesty's ship *Little Belt* was dismasted at Halifax, Nova Scotia, and her bowsprit was found, eighteen months after, in the Basque Roads. The mainmast of the *Tilbury*, buried off Hispaniola in the Seven Years' war, was brought to our shores."*

These facts are dependent on the eastward set of this majestic current; and so is another great physical fact of immeasurable importance to us;— the superiority in temperature of the western shores of Europe over the eastern shores of North America. The harbour of St. John's, Newfoundland, is frequently fast closed by ice in the month of June; yet the latitude of St. John's is considerably south of that of the port of Brest, in France.

Impelled by the rotatory motion of the earth, and by the trade-wind,† the equatorial waters of the Atlantic are ever urged, a broad and rapid river, into the Caribbean sea, and the Gulf of Mexico, the narrowing shores of which compress the stream as in a funnel. The Andes here present a slender but impregnable barrier to its further progress westward; and the trend of the Isthmus

---

* *Blackwood;* April, 1849; p. 412.

† Strictly speaking, the current is a lagging behind of the water, which cannot keep pace with the speed communicated to the solid crust of the globe at its equatorial regions. The trade-wind is owing to the same cause.

turns it to the northward. Still finding no outlet, the impatient current, like a wild-beast pacing round its cage, courses the Gulf of Mexico, doubles the peninsula of Florida, and pursues its way first to the north-east, and then to the east, crossing the Atlantic in a retrograde direction, until it laves with its warm billows the coasts of Europe. Here it turns to the southward, and after embracing the " Fortunate" isles that lie off the African shores,— the Azores, the Madeiras, and the Canaries,—it joins the great equatorial set beneath the trade-wind, and returns on its westward course.

This mighty circulation of water must have been going on from the instant that the earth commenced rotating on its axis, or (granting this to have been chronologically subsequent) from the instant the Atlantic occupied its present bed. Whether sooner or later, it commenced at *some* instant ; but at that instant all the previous elements of the circle were presupposed, and a boundless succession of former circles. An intelligent stranger, looking on the movement immediately after its commencement, but ignorant of its origin, would not be able to assign any limit to its past duration. From his observation of the velocity of the current in different parts of the circle, he would say with confidence,—" These identical particles of water, which I see now urged on their ceaseless course towards the middle of the North Atlantic, were, yesterday morning at this hour, in the latitude of the mouth of the Chesapeake ; on the morning before, off Cape Hatteras ; on the morning before that, off Cape Lookout ;" and so backwards interminably.

Whether the economy of the globe is circular, or not, I am not in a position to show. But its movements certainly are ; and so are the movements of all the myriad worlds with which astronomy is conversant. Asteroids, planets, satellites, comets, suns, —nay, even the stellar universe itself—obey *in their motions*, the grand universal law of circularity. Take any one of these ;—our Moon. When its orbital motion commenced, it commenced at some point or other of the circle which it describes in its course around the earth. The pre-existence, or at least the co-existence, of the Earth, and also that of the Sun, are necessary to its motion. Supposing it possible for a spectator, furnished with modern astronomical knowledge, to have looked at that instant on the newly-spun orb, would he not confidently have inferred, from its position at that moment, its position a week before ? Would he not have felt able to indicate with unhesitating certainty the solar and lunar eclipses of a century or a chiliad before, just as he now calculates the time of the eclipse that marked the death of Herod the Great? Undoubtedly he would; for he would assume the constancy of those movements which modern science has deduced from the observations of many centuries; and, granting him the fact of their constancy, we could not invalidate his conclusions. Yet *what* would he have shown? The conditions and phenomena of bodies before they had begun to exist. The conditions are legitimately deducible; but they are prochronic conditions.

The mention of the celestial orbs suggests to remembrance the famous argument for the vast antiquity of the material universe, founded on the time which is required for the propulsion of light. I believe it owes its origin to Sir William Herschel.

Speaking of the known velocity of light in connexion with the immense distance of certain nebulæ, that eminent astronomer made these remarks:—

"Hence it follows, that, when we . . . see an object of the calculated distance at which one of these very remote nebulæ may still be perceived. . . . . .

the rays of light which convey its image to the eye must have been more than nineteen hundred and ten thousand, that is, almost *two millions,* of years on their way; and that, consequently, so many years ago, this object must already have had an existence in the sidereal heavens, in order to send out those rays by which we now perceive it." *

The notion has been amplified, with some interesting details, by a writer in the *Scottish Congregational Magazine* for *January* 1847; who thus throws the statements into a tabular form, and comments on them.

" From the Moon, light comes to the earth in 1¼ second
,, the Sun ,, ,, in 8 minutes
,, Jupiter ,, ,, in 52 ,,
,, Uranus ,, ,, in 2 hours
,, a fixed Star of 1st magnitude — 3 to 12 years
,, ,, 2d ,, 20 ,,
,, ,, 3d ,, 30 ,,
,, ,, 4th ,, 45 ,,
,, ,, 5th ,, 66 ,,
,, ,, 6th ,, 96 ,,
,, ,, 7th ,, 180 ,,
,, ,, 12th ,, 4000 ,,

" Now, as we see objects by the rays of light passing from those objects to our eye, it follows that we do not perceive the heavenly bodies, *as they*

* Philos. Trans. for 1802; p. 498.

are at the moment of our seeing them, but *as they were* at the time the rays of light by which we see them left those bodies. Thus when we look at the moon, we see her, not as she is at the moment of our beholding her disc, but as she was a second and a quarter before; for instance, we see her not at the moment of her rising above the horizon, but $1\frac{1}{4}$ second after she has risen. The sun also when he appears to us to have just passed the meridian, has already passed it by 8 minutes. So, in like manner, of the planets and the fixed stars. We see Jupiter, not as he is at the moment of our catching a sight of him, but as he was 52 minutes before. Uranus appears to us, not as he is at the moment of our discovering him, but as he was 2 hours previously. And a star of the 12th magnitude presents itself to our eye as it was 4,000 years ago: so that, suppose such a star to have been annihilated 3,000 years back, it would still be visible on the earth's surface for 1,000 years to come: or, suppose a star of the same magnitude had been created at the time the Israelites left Egypt, it will not be perceptible on the earth for nearly 700 years from this date."

Beautiful, and at first sight unanswerable as this argument is, it falls to the ground before the spear-

touch of our Ithuriel, the doctrine of prochronism.
There is nothing more improbable in the notion
that the sensible undulation was created at the
observer's eye, with all the pre-requisite undulations prochronic, than in the notion that blood was
created in the capillaries of the first human body.
The latter we have seen to be a fact : is the former
an impossibility ?

It may perhaps be said :—" The traces of prochronism you have adduced in created organisms
may be granted, because they are inseparable from
the presumed condition of those organisms respectively. The blood in the vessels, the hair, the
teeth, the nails, may afford evidences of past processes ; but then those are only past stages of what
yet exists. The case, however, is not parallel with
the fossil skeletons, many of which have no connexion with anything now existing. The concentric rings of a timber-tree are essential to its
adult state; but how is the existence of the *Pterodactyle* or the *Megatherium* essential to that of the
recent *Draco volans*, or the South American Sloth?
Can you show in the new-formed creature any
trace of some organ which does not come into its
present condition of being,—of something which
has quite passed away ? "

R 2

Perhaps I can. The very concentric rings of the tree are considered by botanists as, in some sense, dead. The paradoxical dictum of Schleiden, —"No tree has leaves," *—is grounded on this circumstance, that the woody portion of the mass is the inert result of former generations, and that the present race of leaves is growing, not out of the woody portion of the tree, but out of its herbaceous extremities, "which grow upon the woody stem *as upon a ground*, formed by the process of vegetation. This common ground, namely, the woody stem, *which is almost lifeless* in comparison with the herbaceous parts engaged in active growth, is annually covered with a vigorous sheath under the protecting bark; and this sheath is the ground of the nourishment of all the vegetating herbaceous extremities." †

The polygonal plates into which the bark of the *Testudinaria* divides, not only show many superposed laminæ, at any given moment of its adult condition, but also bear witness, in the broad existent surface of each one, to former laminæ, yet older than the oldest now present, which have disintegrated and dropped off.

The Palm and the Tree-fern show, in their trunk-

* Beitrage, p. 152.   † Dr. A. Braun, On the Veg. Indiv.

scars, evidences of organs which have completely died away and disappeared; while, between these scars and the generation of living fronds, there is, at any given moment of the tree's history, a series of fronds which are quite dead and dry, but which have not yet disappeared.

The *Nerita,* a genus of beautiful shells from the tropical seas, dissolves away and removes, in the progress of growth, the spiral column, which originally formed the axis of development; so that, in adult age, the spiral direction of the whole testifies to the past existence of a column which has quite disappeared.

In that species of *Murex,** which, on account of the long and slender rostellum, and the spines with which it is covered, is known to collectors as the Thorny Woodcock (*M. tenuispina*), the shelly spines of the earlier whorls would interfere with such as came, in process of development, to be superposed on them; for they cross the area which is to be the cavity enclosed by the advancing lip. They are, however, removed by absorption; but not so completely but that traces may still be discovered where they formerly existed: evidences of the quondam existence of what exists no longer.

* See *ante,* p. 233.

Towards one side of the upper surface of the pretty Star-fish, *Cribella rosea,* (as in many other species of Star-fishes,) there is a curious little mark, known as the *madreporic plate,* the use of which has greatly puzzled naturalists. Sars, the Norwegian zoologist, has unveiled the mystery.* The young larva, before it assumes the stellar form, is furnished with a sort of thick column, divided into four diverging clubbed arms, which are adhering organs, ancillary to locomotion. In the process of development, however, new locomotive organs are formed ; and this four-fold column, being no longer needed, sloughs away; and that so completely, that not a trace of its existence remains, *except this scar,* or " *madreporic plate ;* " which is therefore a permanent record of something that has quite passed away.

But the closest parallel to the relation borne by the skeleton of an extinct species to an extant one, is presented by that of the hilum to a seed, or of the umbilicus to a mammal. Each of these is a legible and undeniable record of a being, whose individuality was totally distinct from that of the being by which it is presented, and of which all vestiges have disappeared, *save this record.* Nor

* Fauna Littor. Norveg.; i. 47.

is the parallel founded on obscure or rare examples ;
both the umbilicus and the hilum are generally
conspicuous; and both are extensively found in
their respective kingdoms, the former pervading
the viviparous Vertebrata, the latter characterising
the whole of the cotyledonous types of vegetation.

Once more. An objection may arise to the reception of the prochronic principle, on the ground
that the examples I have adduced are not to be
compared, in point of grandeur, with the mighty
revolutions, which are presumed to have written
their records in the crust of the globe; and that
hence no analogy can be fairly drawn from one to
the other. To the philosopher, however, there is
no great or small, as there is none in the works of
God. We have every reason to believe that He
has wrought by the same laws in all portions of
his universe: the principle on which an apple falls
from the branch to the ground, is the same as that
which keeps the planet Neptune in the solar system.
I have shown that the principle of prochronic development obtains wherever we are able to test it;
that is, wherever another principle, that of *cyclicism*,
exists ; whether the cycle be that of a gnat's
metamorphosis, or of a planet's orbit. The distinction of great or small, grand or mean, does not

apply to it. If it cannot be proved to be universal, it is only because we are not sufficiently acquainted with some of the economies of nature to be able to pronounce with certainty whether they are cyclical or not. I am not aware of any natural process, in which its existence can be absolutely denied.

And this makes all the difference in the world between my position and that of the old simple-minded observers, with which a superficial reader might think it to possess a good deal in common. A century ago, people used to talk of *lusus naturæ;* of a certain *plastic power* in nature ; of abortive or initiative attempts at making things which were never perfected; of imitations, in one kingdom, of the proper subjects of another, (as plants were supposed to be imitated by the frost on a window-pane, and by the dendritic forms of metals). Still later, many persons have been inclined to take refuge from the conclusions of geology in the absolute sovereignty of God, asking,—" Could not the Omnipotent Creator make the fossils in the strata, just as they now appear? "

It has always been felt to be a sufficient answer to such a demand, that no reason could be adduced for such an exercise of mere power; and that it would be unworthy of the Allwise God.

But this is a totally different thing from that for which I am contending. I am endeavouring to show that a grand LAW exists, by which, in two great departments of nature at least, the analogues of the fossil skeletons were formed without pre-existence. An arbitrary acting, and an acting on fixed and general laws, have nothing in common with each other.

Finally, the acceptance of the principles presented in this volume, even in their fullest extent, would not, in the least degree, affect the study of scientific geology. The character and order of the strata; their disruptions and displacements and injections; the successive floras and faunas; and all the other phenomena, would be *facts* still. They would still be, as now, legitimate subjects of examination and inquiry. I do not know that a single conclusion, now accepted, would need to be given up, except that of actual chronology. And even in respect of this, it would be rather a modification than a relinquishment of what is at present held; we might still speak of the inconceivably long duration of the processes in question, provided we understand *ideal* instead of *actual* time; —that the duration was projected in the mind of God, and not really existent.

The zoologist would still use the fossil forms of non-existing animals, to illustrate the mutual analogies of species and groups. His recognition of their prochronism would in nowise interfere with his endeavours to assign to each its position in the scale of organic being. He would still legitimately treat it as an entity; an essential constituent of the great Plan of Nature; because he would recognise the Plan itself as an entity, though only an ideal entity, existing only in the Divine Conception. He would still use the stony skeletons for the inculcation of lessons on the skill and power of God in creation; and would find them a rich mine of instruction, affording some examples of the adaptation of structure to function, which are not yielded by any extant species. Such are the elongation of the little finger in *Pterodactylus*, for the extension of the alar membrane; and the deflexion of the inferior incisors in *Dinotherium*, for the purposes of digging or anchorage. And still would he find, in the fossil forms, evidences of that complacency in beauty, which has prompted the Adorable Workmaster to paint the rose in blushing hues, and to weave the fine lace of the dragonfly's wing. The whorls of the *Gyroceras*, the foliaceous or zigzag sutures of the *Ammonites*, and the radiating pattern of *Smithia,* are not less elegant than anything of the kind in existing creation, in which, however, they have no parallels. In short, the readings of the "stone book" will be found not less worthy of God who wrote them, not less worthy of man who

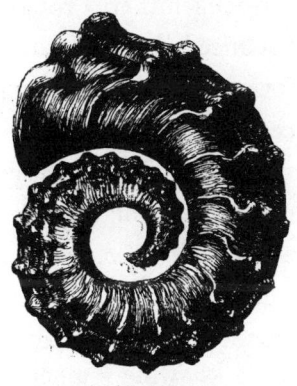

GYROCERAS.

deciphers them, if we consider them as prochronically, than if we judge them diachronically, produced.

—————————

Here I close my labours. How far I have succeeded in accomplishing the task to which I bent myself, it is not for me to judge. Others will determine that; and I am quite sure it will be determined fairly, on the whole. To prevent misapprehension, however, it may be as well to enunciate what the task was, which I prescribed, especially because other (collateral, hypothetical) points have been mooted in these pages.

All, then, that I consider myself responsible for is summed up in these sentences :—

I. The conclusions hitherto received have been but inferences deduced from certain premises: the witness who reveals the premises does *not* testify to the inferences.

II. The process of deducing the inferences has been liable to a vast incoming of error, arising from the operation of a *Law*, proved to exist, but hitherto unrecognised.

III. The amount of the error thus produced we have no means of knowing; much less of eliminating it.

IV. The whole of the facts deposed to by this witness are irrelevant to the question; and the witness is, therefore, out of court.

V. The field is left clear and undisputed for the one Witness on the opposite side, whose testimony is as follows :—

" IN SIX DAYS JEHOVAH MADE HEAVEN AND EARTH, THE SEA, AND ALL THAT IN THEM IS."

INDEX.

R. CLAY, PRINTER, BREAD STREET HILL.

MARINE
NATURAL HISTORY CLASS.

In the summer of 1855, I met, at Ilfracombe, on the coast of North Devon, a small party of ladies and gentlemen, who formed themselves into a Class for the study of Marine Natural History. There was much to be done in the way of collecting, much to be learned in the way of study. Not a few species of interest, and some rarities, fell under our notice, scattered as we were over the rocks, and peeping into the pools, almost every day for a month. Then the prizes were to be brought home, and kept in little Aquariums for the study of their habits, their beauties to be investigated by the pocket-lens, and the minuter kinds to be examined under the microscope. An hour or two was spent on the shore every day on which the tide and the weather were suitable; and, when otherwise, the occupation was varied by an indoors' lesson, on identifying and comparing the characters of the animals obtained, the specimens themselves affording illustrations. Thus the two great desiderata of young naturalists were attained simultaneously; they learned at the same time how to collect, and how to determine the names and the zoological relations of the specimens when found.

A little also was effected in the way of dredging the sea-bottom, and in surface-fishing for Medusæ, &c.; but our chief attention was directed to shore-collecting. Altogether, the experiment was found so agreeable, that I propose to repeat it by forming a similar party every year, if spared, at some suitable part of the coast.

Such ladies or gentlemen as may wish to join the Class should give in their names to me, early in the summer; and any preliminary inquiries about plans, terms, &c. shall meet the requisite attention.

P. H. GOSSE.

Marychurch, Torquay,
Oct. 1857.

Early in 1858, (D.V.) *will be published, the First Number*

A HISTORY

OF THE

BRITISH SEA-ANEMONES,

BY P. H. GOSSE, F.R.S.

In bi-monthly Numbers, each containing 32 pp. 8vo. and a coloured
plate.   Price 1*s*. 6*d*.

———————

MR. GOSSE has for some years been collecting materials for a
complete history of our native Sea-Anemones, with illustrations of
every species, drawn and coloured by himself from living specimens.

In order to further this project, which is now in immediate prospect,
he respectfully invites the co-operation of his kind scientific friends at
various parts of the British and Irish Coasts, who may materially assist
him by transmitting to him specimens of all species that are not
common everywhere.

An Anemone of medium size may be safely sent *by post*, in a small
tin-canister, *without water*, but with a small tuft of damp sea-weed, rag,
or blotting-paper, to maintain a moist atmosphere around the animal.
A piece of paper should be *pasted* round the canister, to secure it, and
also to receive the address; and the whole would probably come within
the weight covered by a twopenny or fourpenny stamp.   It is important that no rattling of water be audible, and that no exudation take
place ; as in either case, the package would be detained at the Post-
office.

MARYCHURCH, TORQUAY,
October, 1857.

WORKS ON MARINE NATURAL HISTORY AND THE AQUARIUM;

BY PHILIP HENRY GOSSE, F.R.S.

A NATURALIST'S RAMBLES

ON

THE DEVONSHIRE COAST.

With TWENTY-EIGHT PLATES, some coloured. Post 8vo. 21s.

" The charming book now before us . . . . The lively pages of this graphic and well-illustrated volume . . . . We know of no book where that beautiful family, the Sea-Anemones, are more graphically described and brought before the eye of the reader."—*Fraser's Magazine*, Oct. 1853.

"This charming volume, which we so strongly recommend to our readers . . . . largely enters into the private history [of the Sea-Anemones and other Zoophytes], and to the attractions of an engaging style and healthy piety, adds the accompaniment of elaborately coloured drawings of the animals themselves."—*Leisure Hour*, Feb. 9, 1854.

"Scarcely have we pronounced a most favourable opinion of Mr. Gosse's ' Naturalist's Sojourn in Jamaica,' than we are called upon to review another book from the same pen, equally beautiful, equally amusing, and equally instructive . . . . . This is a fit companion to the ' Sojourn;' like that, it is a series of pictures which it must delight the lover of nature to look upon . . . . the animals of the sea are here revealed to us in all their most attractive forms."—*Zoologist*, Oct. 1853.

" The present will ably support the previous character of its talented author."—*Natural History Review*, Jan. 1854.

LONDON: JOHN VAN VOORST, PATERNOSTER ROW.

WORKS BY PHILIP HENRY GOSSE, F.R.S.

A MANUAL OF MARINE ZOOLOGY

FOR

THE BRITISH ISLES.

Two Vols. Foolscap 8vo. with nearly 700 Engravings, 15s.

THIS Work gives in plain English terms the characters by which to determine the Class, Order, Family, and Genus of *every animal* known to inhabit the British Seas. Every Genus is illustrated by a figure, drawn by the Author, principally from nature, and is accompanied by a list of the recognised Species.

A need long felt is supplied by this book, which, it is hoped, will be found a valuable *vade mecum*, if not indispensable, to every visitor to the sea-side, who desires acquaintance with its living treasures.

Every Class is introduced by a *résumé* of the most interesting points of its Natural History, with notes of the localities frequented by the Species, and directions for identifying them.

PART I.

I. SPONGES.	V. STARFISHES.	IX. CRUSTACEA.
II. FORAMINIFERA.	VI. TURBELLARIA.	X. CIRRIPEDIA.
III. ZOOPHYTES.	VII. ANNELIDA.	XI. MITES.
IV. MEDUSÆ.	VIII. ROTIFERA.	XII. INSECTS.

PART II.

XIII. POLYZOA.	XVI. BRACHIOPODA.	XIX. CEPHALOPODA.
XIV. TUNICATA.	XVII. PTEROPODA.	XX. FISHES.
XV. CONCHIFERA.	XVIII. GASTROPODA.	XXI. MAMMALIA.

LONDON: JOHN VAN VOORST, PATERNOSTER ROW.

WORKS BY PHILIP HENRY GOSSE, F.R.S.

_____

T E N B Y:

A S E A - S I D E H O L I D A Y.

With 24 Plates, coloured, post 8vo. 21s.

_____

" Here we have another issue from the fertile pen of Mr. Gosse, and another of his delightful sea-side books. It is fully worthy of its predecessors in pleasant gossip, in interesting information, in important scientific novelty, and in variety and beauty of illustration."—_Athenæum_, May 31, 1856.

" It is the history of a month spent by a man of research, in the pursuit of a favourite study, under favourable circumstances; and is full of original investigations, successful observations, and pleasing descriptions of the impressions produced by novel objects upon an unaffected and healthy mind. It is a book we cannot read without regretting, as we pass from page to page with increasing interest, that we were not his companions . . . . No. intelligent reader can rise from the perusal of ' Tenby ' without gaining much knowledge from a delightful book."—_Eclectic Review_, June, 1856.

" Mr. Gosse tells us how he got to Tenby; talks of the places there, the caverns, Monkstone, North Cove, Hean Castle, Hoyle's Mouth, Tenby Head, and other places to be visited; shows where the marine animals, his favourites, most abound; teaches how to get at them, when to catch them in a visible condition, how to keep them, how to study them, and what their points of interest are. Of such matters is the book made up, and to us it seems to be perfect in its way."—_Gardener's Chronicle_, May 17, 1856.

" The natural history is admirable, the descriptions picturesque and vivid in a very uncommon degree, and the illustrations excellent. Mr. Gosse has, in his various books, added a great deal to our knowledge of marine [animals], many of them microscopic; and this book is amongst his best on this subject."—_Guardian_, June 11, 1856.

" This charming issue from his fertile pen will delight scores of naturalists, as well as induce a liking for a healthy and rational amusement among the many loungers who indulge in a sea-side holiday."—_Lincolnshire Times_, June 10, 1856.

_____

LONDON: JOHN VAN VOORST, PATERNOSTER ROW.